OPTIMAL CONTROL MODELS IN FINANCE

Applied Optimization
Volume 95

Series Editors:

Panos M. Pardalos
University of Florida, U.S.A.

Donald W. Hearn
University of Florida, U.S.A.

OPTIMAL CONTROL MODELS IN FINANCE
A New Computational Approach

by

PING CHEN
Victoria University, Melbourne, Australia

SARDAR M.N. ISLAM
Victoria University, Melbourne, Australia

 Springer

Library of Congress Cataloging-in-Publication Data

A C.I.P. Catalogue record for this book is available
from the Library of Congress.

ISBN 0-387-23569-8   e-ISBN 0-387-23570-1   Printed on acid-free paper.

© 2005 Springer Science+Business Media, Inc.
All rights reserved. This work may not be translated or copied in whole or in part without
the written permission of the publisher (Springer Science+Business Media, Inc., 233 Spring
Street, New York, NY 10013, USA), except for brief excerpts in connection with reviews or
scholarly analysis. Use in connection with any form of information storage and retrieval,
electronic adaptation, computer software, or by similar or dissimilar methodology now
know or hereafter developed is forbidden.
The use in this publication of trade names, trademarks, service marks and similar terms,
even if the are not identified as such, is not to be taken as an expression of opinion as to
whether or not they are subject to proprietary rights.

Printed in the United States of America.

9 8 7 6 5 4 3 2 1   SPIN 1161547

springeronline.com

Contents

List of Figures		ix
List of Tables		xi
Preface		xiii
Introduction		xv

1. **OPTIMAL CONTROL MODELS** — 1
   1. An Optimal Control Model of Finance — 2
   2. (Karush) Kuhn-Tucker Condition — 4
   3. Pontryagin Theorem — 6
   4. Bang-Bang Control — 7
   5. Singular Arc — 7
   6. Indifference Principle — 8
   7. Different Approaches to Optimal Control Problems — 10
   8. Conclusion — 20

2. **THE STV APPROACH TO FINANCIAL OPTIMAL CONTROL MODELS** — 21
   1. Introduction — 21
   2. Piecewise-linear Transformation — 21
   3. Non-linear Time Scale Transformation — 23
   4. A Computer Software Package Used in this Study — 25
   5. An Optimal Control Problem When the Control can only Take the Value 0 or 1 — 26
   6. Approaches to Bang-Bang Optimal Control with a Cost of Changing Control — 27
   7. An Investment Planning Model and Results — 30

	8	Financial Implications and Conclusion	36
3.	A FINANCIAL OSCILLATOR MODEL		39
	1	Introduction	39
	2	Controlling a Damped Oscillator in a Financial Model	40
	3	Oscillator Transformation of the Financial Model	41
	4	Computational Algorithm: The Steps	44
	5	Financial Control Pattern	47
	6	Computing the Financial Model: Results and Analysis	47
	7	Financial Investment Implications and Conclusion	89
4.	AN OPTIMAL CORPORATE FINANCING MODEL		91
	1	Introduction	91
	2	Problem Description	91
	3	Analytical Solution	94
	4	Penalty Terms	98
	5	Transformations for the Computer Software Package for the Finance Model	99
	6	Computational Algorithms for the Non-linear Optimal Control Problem	101
	7	Computing Results and Conclusion	104
	8	Optimal Financing Implications	107
	9	Conclusion	108
5.	FURTHER COMPUTATIONAL EXPERIMENTS AND RESULTS		109
	1	Introduction	109
	2	Different Fitting Functions	109
	3	The Financial Oscillator Model when the Control Takes Three Values	120
	4	Conclusion	139
6.	CONCLUSION		141
Appendices			144
A	CSTVA Program List		145
	1	Program A: Investment Model in Chapter 2	145
	2	Program B: Financial Oscillator Model in Chapter 3	149
	3	Program C: Optimal Financing Model in Chapter 4	153
	4	Program D: Three Value-Control Model in Chapter 5	156

Contents vii

B	Some Computation Results		161
	1	Results for Program A	161
	2	Results for Program B	163
	3	Results for Program C	167
	4	Results for Program D	175
C	Differential Equation Solver from the SCOM Package		181
D	SCOM Package		183
E	Format of Problem Optimization		189
F	A Sample Test Problem		191

References 193

Index 199

List of Figures

2.1	Plot of n=2, forcing function ut=1,0	31
2.2	Plot of n=4, forcing function ut=1,0,1,0	31
2.3	Plot of n=6, forcing function ut=1,0,1,0,1,0	32
2.4	Plot of n=8, forcing function ut=1,0,1,0,1,0,1,0	32
2.5	Plot of n=10, forcing function ut=1,0,1,0,1,0,1,0,1,0	33
2.6	Plot of the values of the objective function to the number of the switching times	34
2.7	Plot of the cost function to the cost of switching control	35
3.1	Plot of integral F against 1/ns at ut=-2,2	87
3.2	Plot of integral F against 1/ns at ut=2,-2	87
3.3	Plot of cost function F against the number of large time intervals nb	88
5.1	Plot of n=4, forcing function ut=1,0,1,0	110
5.2	Plot of n=10, forcing function ut=1,0,1,0,1,0,1,0,1,0	112
5.3	Results of objective function at n=2,4,6,8,10	113
5.4	Plot of n=4, forcing function ut=1,0,1,0	116
5.5	Plot of n=8, forcing function ut=1,0,1,0,1,0,1,0	117
5.6	Plot of n=8, forcing function ut=1,0,1,0,1,0,1,0	119
5.7	Plot of nb=9, ns=8, forcing function ut=-2,0,2,-2,0,2,-2,0,2	123
5.8	Relationship between two state functions during the time period 1,0	123

List of Tables

2.1	Objective functions with the number of the switching times	33
2.2	Costs of the switching control attached to the objective function	34
3.1	Results of the objective function at control pattern -2,2, ...	48
3.2	Results of the objective function at control pattern 2,-2, ...	86
4.1	Computing results for solution case [1]	105
4.2	Computing results for solution case [2]	106
4.3	Computing results for solution case 2 with another mapping control	106
5.1	Results of objective function at n=2,4,6,8,10	114
5.2	Results of objective functions at n=2,6,10	118
5.3	Test results of the five methods	120
5.4	Results of financial oscillator model	121

Preface

This book reports initial efforts in providing some useful extensions in financial modeling; further work is necessary to complete the research agenda. The demonstrated extensions in this book in the computation and modeling of optimal control in finance have shown the need and potential for further areas of study in financial modeling. Potentials are in both the mathematical structure and computational aspects of dynamic optimization. There are needs for more organized and coordinated computational approaches. These extensions will make dynamic financial optimization models relatively more stable for applications to academic and practical exercises in the areas of financial optimization, forecasting, planning and optimal social choice.

This book will be useful to graduate students and academics in finance, mathematical economics, operations research and computer science. Professional practitioners in the above areas will find the book interesting and informative.

The authors thank Professor B.D. Craven for providing extensive guidance and assistance in undertaking this research. This work owes significantly to him, which will be evident throughout the whole book. The differential equation solver "nqq" used in this book was first developed by Professor Craven. Editorial assistance provided by Matthew Clarke, Margarita Kumnick and Tom Lun is also highly appreciated. Ping Chen also wants to thank her parents for their constant support and love during the past four years.

PING CHEN AND SARDAR M.N. ISLAM

Introduction

Optimal control methods have significant applications in finance. This book discusses the general applications of optimal control methods to several areas in finance with a particular focus on the application of bang-bang control to financial modeling.

During the past half-century, many optimization problems have arisen in fields such as finance management, engineering, computer science, production, industry, and economics. Often one needs to optimize (minimize or maximize) certain objectives subject to some constraints. For example, a public utility company must decide what proportion of its earnings to retain to the advantage of its future earnings at the expense of gaining present dividends, and also decide what new stock issues should be made. The objective of the utility is to maximize the present value of share ownership, however, the retention of retained earnings reduces current dividends and new stock issues can dilute owners' equity.

Some optimization problems involve optimal control, which are considerably more complex and involve a dynamic system. There are very few real-world optimal control problems that lend themselves to analytical solutions. As a result, using numerical algorithms to solve the optimal control problems becomes a common approach that has attracted attention of many researchers, engineers and managers. There are many computational methods and theoretical results that can be used to solve very complex optimal control problems. So computer software packages of certain optimal control problems are becoming more and more popular in the era of a rapidly developing computer industry. They rescue scientists from large calculations by hand.

Many real-world financial problems are too complex for analytical solutions, so must be computed. This book studies a class of optimal financial control problems where the control takes only two (or three) different discrete values. The non-singular optimal control solution of linear-analytic systems in finance with bounded control is commonly known as the bang-bang control. The problem of finding the optimal control becomes one of finding the switching times in the dynamic financial system. A cost of switching control is added to usual models since there is a cost for switching from one financial instrument to another. Computational algorithms based on the time scaled transformation technique are developed for this kind of problems. A set of computer software packages named CSTVA is generated for real-world financial decision-making models.

The focus in this research is the development of computational algorithms to solve a class of non-linear optimal control problems in finance (bang-bang control) that arise in operations research. The Pontryagin theory [69, 1962] of optimal control requires modification when a positive cost is associated with each switching of the control. The modified theory, which was first introduced by Blatt [2, 1976], will give the solutions of a large class of optimal control problems that cannot be solved by standard optimal control theories. The theorem is introduced but not used to solve the problems in this book. However, the cost of changing control, which is attached to the cost function, is used here for reaching the optimal solution in control system. In optimization computation, especially when calculating minimization of an integral, an improved result can be obtained by using a greater number of time intervals.

In this research, a modified version of the Pontryagin Principle, in which a positive cost is attached to each switching of the control, indicates that a form of bang-bang control is optimal. Several computational algorithms were developed for such financial control problems, where it is essential to compute the switching times. In order to achieve the possibility of computation, some transformations are included to convert control functions, state functions and the integrals from their original mathematical forms to computable forms. Mainly, the MATLAB "constr" optimization package was applied to construct the general computer programs for different classes of optimal control problems. A simplified financial optimal control problem that only has one state and one control is introduced first. The optimal control of such a problem is bang-bang control, which switches between two values in successive time intervals. A computer software package was developed for solving this particular problem, and accurate results were obtained. Also some transformations are applied into the problem formalization. A financial oscillator problem is then treated, which has two states and one control. The transformation of subdivision of time interval technique is used to gain a more accurate gradient. Different sequences of control are then studied. The computational algorithms are applied to a non-linear optimal control problem of an optimal financing model, which was original introduced by Davis and Elzinga [22, 1970]. In that paper, Davis and Elzinga had an analytical solution for the model. In this book a computer software package was developed for the same model, including setting up all the parameters, calculating the results, and testing different initial points of an iterative algorithm. During the examination of the algorithms, it

INTRODUCTION

was found that sometimes a local minimum was reached instead of a global optimum. The reasons for the algorithms leading to such a local minimum are indicated, and as a result, a part of the algorithms are modified so as to obtain the global optimum eventually.

The computing results were obtained, and are presented in graphical forms for future analysis and improvement here. This work is also compared with other contemporary research. The advantages and disadvantages of them are analyzed. The STV approach provides an improved computational approach by combing the time discretization method, the control step function method, the time variable method, the consideration of transaction costs and by coding the computational requirements in a widely used programming system MATLAB. The computational experiments validated the STV approach in terms of computational efficiency, and time, and the plausibility of results for financial analysis.

The present book also provides a unique example of the feasibility of modeling and computation of the financial system based on bang-bang control methods. The computed results provide useful information about the dynamics of the financial system, the impact of switching times, the role of transaction costs, and the strategy for achieving a global optimum in a financial system.

One of the areas of applications of optimal control models is normative social choice for optimal financial decision making. The optimal control models in this book have this application as well. These models specify the welfare maximizing financial resource allocation in the economy subject to the underlying dynamic financial system.

Chapter 1 is an introduction to the optimal control problems in finance and the classical optimal control theories, which have been successfully used for years. Some relevant sources in this research field are also introduced and discussed.

Chapter 2 discusses a particular case of optimal control problems and the switching time variable (STV) algorithm. Some useful transformations introduced in Section 2.2 are standard for the control problems. The piecewise-linear transformation and the computational algorithms discussed in Section 2.6 are the main work in this book. A simple optimal aggregate investment planning model is presented here. Accurate results were obtained in using these computational algorithms, and are presented in Section 2.7. A part of the computer software SCOM developed in Craven and Islam [18, 2001] and Islam and Craven [38, 2002] is used here to solve the differential equation.

Chapter 3 presents a financial oscillator model (which is a different version of the optimal aggregative investment planning model developed in Chapter 2) whose state is a second-order differential equation. A new time-scaled transformation is introduced in Section 3.3. The new transformation modifies the old transformations that are used in Chapter 2. All the modifications are made to match the new time-scale division. The computational algorithms for this problem and the computing results are also discussed. An extension of the control pattern is indicated. The new transformation and algorithms in this chapter are the important parts in this research.

Chapter 4 contains an optimal financing model, which was first introduced by Davis and Elzinger [22, 1970]. A computer software package for this model is constructed in this book (for details see Appendix A.3 model1_1.m - model1_5.m). The computing result is compared with the analytical result and another computing result obtained from using the SCOM package.

Chapter 5 reports computing results of the algorithms 2.1-2.3 and algorithms 3.1-3.4 in other cases of optimal control problems. After analyzing the results, the computer packages in Appendix A.1 and Appendix A.2 (project1_1.m - project1_4.m and project2_2.m - project2_4.m) have been improved.

Chapter 6 gives the conclusion of this research. Optimal control methods have high potential applications to various areas in finance. The present study has enhanced the state of the art for applying optimal control methods, especially the bang-bang control method, for financial modeling in a real life context.

Chapter 1

OPTIMAL CONTROL MODELS IN FINANCE

Optimal control theory has been important in finance (Islam and Craven [38, 2002]; Taperio [81, 1998]; Ziemba and Vickson [89, 1975]; Senqupta and Fanchon [80, 1997]; Sethi and Thompson [79, 2000]; Campbell, Lo and MacKinlay [7, 1997]; Eatwell, Milgate and Neuman [25, 1989]). During nearly fifty years of development and extension of optimal control theories, they have been successfully used in finance. Many famous models effectively utilize optimal control theories. However, with the increasing requirements of more workable and accurate solutions to optimal control problems, there are many real-world problems which are too complex to lead to analytical solutions. Computational algorithms therefore become essential tools for most optimal control problems including dynamic optimization models in finance.

Optimal control modeling, both deterministic and stochastic, is probably one of the most crucial areas in finance given the time series characteristics of financial systems' behavior. It is also a fast growing area of sophisticated academic interest as well as practice using analytical as well as computational techniques. However, there are some limits in some areas in the existing literature in which improvements are needed. It will facilitate the discipline if dynamic optimization in finance is to be at the same level of development in modeling as the modeling of optimal economic growth (Islam [36, 2001]; Chakravarty [9, 1969]; Leonard and Long [52, 1992]). These areas are: (a) specification of the element of the dynamic optimization models; (b) the structure of the dynamic financial system; (c) mathematical structure; and (d) computational methods and programs. While Islam and Craven [38, 2002] have recently made some extensions to these areas, their work does not explicitly focus on bang-bang control models in finance. The objective of this book is to present some suggested improvements in modeling bang-bang control in finance in the deterministic optimization strand by extending the existing literature.

In this chapter, a typical general financial optimal control model is given in Section 1.1 to explain the formula of the optimal control problems and their accompanying optimal control theories. In addition, some classical concepts in operations research and famous standard optimal control theories are introduced in Section 1.2-1.5, and a brief description on how they are applied in financial optimal control problems is also discussed. In Section 1.6, some improvements that are needed to meet the higher requirements for the complex real-world problems are presented. In Section 1.7, the algorithms on similar optimal control problems achieved by other researchers are discussed. Critical comparisons of the methods used in this research and those employed in others' work are made, and the advantages and disadvantages between them are shown to motivate the present research work.

1. An Optimal Control Model of Finance

Consider a financial optimal control model:

$$\text{MIN}_{x(.),u(.)} J(u) := \int_0^T f(x(t), u(t), t) dt \quad (1.1)$$

subject to:

$$x(0) = x_0, \quad \dot{x}(t) = m(x(t), u(t), t) \quad (1.2)$$

$$a \leq u(t) \leq b \quad (1.3)$$

$$0 \leq t \leq T, T > 0 \quad (1.4)$$

Here $x(.)$ is the *state*, and $u(.)$ is the *control*. The time T is the "planning horizon". The differential equation ($\dot{x}(t) = ...$) describes the dynamics of the financial system; it determines $x(.)$ from $u(.)$. It is required to find an optimal $(\bar{x}, \bar{u})$ which minimizes $J(.)$. (We may consider $J(.)$ as a *cost* function.) Although the control problem is stated here (and other chapters in this book) as a minimization problem, many financial optimization models are in a maximization form. Detailed discussion of control theory applications to finance may be seen in Sethi and Thompson [79, 2000].

Often, $u(.)$ is taken as a piece-wise continuous function (note that jumps are always needed if the problem is linear in the control $u(.)$, to reach an optimum), and then $x(.)$ is a piece-wise smooth function. A financial optimal control model is represented by the formula (1.1)-(1.4), and can always use the "maximum principle" in Pontryagin theory [69, 1962]. The cost function is usually the sum of the integral cost and terminal cost in a standard optimal control problem which can be found in references [1, 1988], [5, 1975], and [8, 1983]. However, for a large class of reasonable cases, there are often no available results from standard theories. A more acceptable method is needed. In Blatt [2, 1976], there is some cost added when switching the control. The cost can be wear and tear on the switching mechanism, or it can be as the cost of "loss of confidence in a stop-go national economy". The cost is associated with each switching of control. The optimal control problem in financial decision making with a cost of switching control can be described as follows:

$$\text{MIN} J = J(u) + k\gamma = \int_0^T f(x(t), v(t), t)dt + k\gamma \quad (1.5)$$

subject to:

$$x(0) = x_0, \dot{x}(t) = m(x(t), u(t), t) \quad (1.6)$$

$$u(t) = \text{piece-wise constant with at most } N \text{ jumps} \quad (1.7)$$

$$0 \leq t \leq T, T > 0 \quad (1.8)$$

$$k \leq N, \text{ the control jumps} \quad (1.9)$$

Here, γ is the positive cost. k is an integer, representing the number of times the control jumps during the planning period $[0, T]$. In particular, u may be a piece-wise constant, thus it can be approximated by a step-function. Only fixed-time optimal control problems are considered in this book which means T is a constant. Although time-optimal control problems are also very interesting, they have not been considered in this research.

The essential elements of an optimal control model (see Islam [36, 2001]) are: (i) an optimization criteria; (ii) the form of inter-temporal time preference or discounting; (iii) the structure of dynamic systems-under modeling; and(iv) the initial and terminal conditions. Although the literature on the methodologies and practices in the specification of the elements of an optimal control model is well developed in various areas in economics (such as optimal growth, see Islam [36, 2001]; Chakravarty [9, 1969]), the literature in finance in this area is not fully developed (see dynamic optimization modeling in Taperio [81, 1998]; Sengupta and Fanchon [80, 1997]; Zieamba and Vickson [89, 1975]). The rationale for and requirements of the specification of the above four elements of dynamic optimization financial models are not provided in the existing literature except in Islam and Craven [38, 2002]. In the present study, the main stream practices are adopted: (i) an optimization criteria (of different types in different models); (ii) the form of inter-temporal time preference-positive or zero discounting; (iii) the structure of dynamic systems under modeling (different types — linear and non-linear); and (iv) the initial and terminal conditions — various types in different models.

Optimal control models in finance can take different forms including the following: bang-bang control, deterministic and stochastic models, finite and infinite horizon models, aggregative and disaggregative, closed and open loop models, overtaking or multi-criteria models, time optimal models, overlapping generation models, etc.(see Islam [36, 2001]).

Islam and Craven [38, 2002] have proposed some extensions to the methodology of dynamic optimization in finance. The proposed extensions in the computation and modeling of optimal control in finance have shown the need and potential for further areas of study in financial modeling. Potentials are in both mathematical structure and computational aspects of dynamic optimization. These extensions will make dynamic financial optimization models relatively more organized and coordinated. These extensions have potential applications for academic and practical exercises. This book reports initial efforts in providing some useful extensions; further work is necessary to complete the research agenda.

Optimal control models have applications to a wide range of different areas in finance: optimal portfolio choice, optimal corporate finance, financial engineering, stochastic finance, valuation, optimal consumption and investment, financial planning, risk management, cash management, etc. (Tapiero [81, 1998]; Sengupta and Fanchon [80, 1997]; Ziemba and Vickson [89, 1975]).

As it is difficult to cover all these applications in one volume, two important areas in financial applications of optimal control models — optimal investment planning for the economy and optimal corporate financing — are considered in this book.

2. (Karush) Kuhn-Tucker Condition

The Kuhn-Tucker Condition condition is the necessary condition for a local minimum of a minimization problem (see [14, 1995]). The Hamiltonian of the Pontryagin maximum principle is based on a similar idea that deals with optimal control problems.

Consider a general mathematical programming problem:

$$\text{MIN} f(x)$$

subject to:

$$g_1(x) \leq 0 \quad (1.10)$$

$$g_2(x) \leq 0 \quad (1.11)$$

$$\ldots$$

$$g_m(x) \leq 0 \tag{1.12}$$
$$h_1(x) = 0 \tag{1.13}$$
$$h_2(x) = 0 \tag{1.14}$$
$$\ldots$$
$$h_r(x) = 0 \tag{1.15}$$
$$x \in R^n \tag{1.16}$$

The Lagrangian is:

$$\begin{aligned} L(x) &= f(x) + \lambda_1 g_1(x) + \lambda_2 g_2(x) + \ldots + \lambda_m g_m(x) \\ &\quad + \mu_1 h_1(x) + \mu_2 h_2(x) + \ldots + \mu_r h_r(x) \end{aligned}$$

The **(Karush-) Kuhn-Tucker** conditions necessary for a (local) minimum of the problem at $x = x^*$ are that Lagrange multipliers λ_i and μ_j exist, for which x^* satisfies the constraints of the problem, and:

$$\nabla L(x^*) = 0; \tag{1.17}$$
$$\lambda_1 \geq 0, \lambda_2 \geq 0, \ldots, \lambda_m \geq 0; \tag{1.18}$$
$$\lambda_1 g_1(x^*) = 0, \mu_1 h_1(x) = 0, \ldots, \lambda_m g_m(x^*) = 0, \mu_m h_m = 0, \ldots; \tag{1.19}$$

The inequality constraints are written here as ≤ 0, then the corresponding $\lambda_i \geq 0$ at the minimum; the multipliers of the equality constraints can take any sign. So for some minimization problems where the inequality constraints are represented as ≥ 0, the sign of the inequalities should be changed first. Then the KKT condition can be applied.

The conditions for global minimum are that the objective and constraint functions are differentiable, and satisfy the constraint qualifications and are convex and concave respectively. If these functions are strictly convex then the minimum is also a unique minimum. Further extensions to these conditions have been made in the cases when the objective and constraint functions are quasi-convex and invex (see Islam and Craven [37, 2001]).

Duality properties of the programming models in finance not only provide useful information for computing purposes, but also for determining efficiency or show prices of financial instruments.

3. Pontryagin Theorem

The Pontryagin theorem was first introduced in Pontryagin [69, 1962]. Consider a minimization problem in finance given as follows:

$$\text{MIN}_{x \in x_0 + X, u \in U} J = \int_0^T f(x(t), u(t), t) dt \qquad (1.20)$$

T is planning horizon, subject to a differential equation and some constraint:

$$\dot{x}(t) = m(x(t), u(t), t) \qquad (1.21)$$

$$g(u) \in K_u \qquad (1.22)$$

Here (1.23) represents the differential equation. (1.24) represents the constraint on control u.

Let the optimal control problem (1.22) reach a (local) minimum at $(\bar{x}, \bar{u})$ with respect to the L_1-norm for u. Assume that f and m are partially Fréchet differentiable with respect to x, uniformly in u near $\bar{u}$. The Hamiltonian is shown as follows:

$$H(\lambda, x, u, t) = -f(x(t), u(t), t) + \lambda(t) m(x(t), u(t), t) \qquad (1.23)$$

The necessary conditions for the minimum are:
(a) a co-state function $\lambda(.)$ satisfies the adjoint differential equation:

$$-\frac{d\lambda}{dt} = f_x(\bar{x}(t), \bar{u}(t), t) - \lambda(t) m_x(\bar{x}(t), \bar{u}(t), t) \qquad (1.24)$$

$$\lambda(T) = 0 \qquad (1.25)$$

with boundary condition.
(b) the associated problems minimize at $\bar{u}(t)$, for all t except at a null set.

4. Bang-Bang Control

In some optimal control problems, when the dynamic equation is linear in the control $u(t)$, bang-bang control (the control only jumps on the extreme points in the feasible area of the constraints on the control) is likely to be optimal. Here a small example is used to explain this concept. Consider the following constraints on the control:

$$0 \leq u_1(t) + u_2(t) \leq 1$$

$$u_1(t) \geq 0, u_2(t) \geq 0$$

In this case the control $u(t) = (u_1(t), u_2(t))$ is restricted to the area of a triangle. The control, which is only staying on the vertices of the triangle and also jumping from one vertex to the other in successive switching time intervals, is the optimal solution. This kind of optimal control is called bang-bang control. The concept can also be used to explain a control:

$$a \leq u(t) \leq b$$

where the optimal control $u(t)$ only takes two possible cases a or b depending on the initial value of the control. This control is also a bang-bang control.

In this research, only bang-bang optimal control models in finance are considered. Sometimes, a singular arc (see Section 1.5) might occur following a bang-bang control in a particular situation. So the possibility of a singular arc occurring should always be considered after a bang-bang control is obtained in the first stage.

5. Singular Arc

As mentioned earlier, when the objective function and dynamic equation are linear in the control, a singular arc might occur following the bang-bang control. In that case, the co-efficient of control $u(t)$ in the associated problem equals zero, thus the control equals zero. So after discovering a bang-bang control solution, it is necessary to check whether a singular arc exists.

In what kind of situation will the singular arc occur? Only when the co-efficient of $u(t)$ in the associated problem may happen to be identically zero for some time interval of time t, say $\gamma \leq t \leq \delta$. The optimum path for such an interval (γ, δ) is called a singular arc; on such an arc, the associated problem only gives $\dot{\lambda}(t) \doteq 0$. A singular arc is very common in real-world trajectory following problems.

6. Indifference Principle

In Blatt [2, 1976], certain financial optimal control models that are concerned with optimal control with a cost of switching control were discussed, and an optimal policy was proved to exist. Also, the maximum principle of the Pontryagin theory was replaced by a weaker condition theorem ("indifference principle"), and several new theories were developed for solving the optimal control problems with cost of changing control. This research is dealing with an optimal control problem with a cost of changing control. Although the "indifference principle" theory is not being employed for solving the optimal control problems in this study, it is still very important to be introduced for understanding the ideas of this research.

Consider a financial optimal control model as follows:

$$\text{MIN } J = x_0(T) + k\gamma = \int_0^T f_0[x(t), v(t), t]dt + k\gamma \qquad (1.26)$$

subject to:

$$\frac{dx}{dt} = f[x(t), v(t), t] \qquad (1.27)$$

$$v(t) = \frac{1}{2} + (-1)^m [v(0) - \frac{1}{2}] \text{ for } t_m < t \le t_{m+1} \text{ and } m = 0, 1, 2, \ldots, k \qquad (1.28)$$

$$f_0(x, v, t) \ge 0 \text{ everywhere inside } R \qquad (1.29)$$

Let:

$$P = (v(0) \, ; \, k \, ; \, t_1, t_2, \ldots, t_k) \qquad (1.30)$$

where $v(0) = 0$ or 1 is the control setting at time $t = 0$; the non-negative integer k is the number of times control alters during time horizon T; and the $t_1, t_2, \ldots, t_k$ are the times at which control alters, satisfying:

$$t_0 \equiv 0 < t_1 < t_2 < \ldots < t_k < T \equiv t_{k+1} \qquad (1.31)$$

Optimal Control Models

Given the policy P, the control function $u(t)$ is shown in (1.30). Now the Hamiltonian of the Pontryagin theory is constructed as follows:

$$H(\lambda, x, v, t) = \lambda(t)f(x, v, t) - f_0(x, v, t) \tag{1.32}$$

The co-state equation is:

$$\frac{d\lambda}{dt} = -\frac{\partial H}{\partial x} = \frac{\partial f_0}{\partial x} - \lambda \frac{\partial f}{\partial x} \tag{1.33}$$

The end-point condition:

$$\lambda(T) = 0 \tag{1.34}$$

Theorem 2. An admissible optimal policy exists. See proof in reference [2, 1976].

Theorem 5. The indifference principle:

Let $P(1.32)$ be an optimal policy. Let H and $\lambda(t)$ be defined by (1.34), (1.35) and (1.36). Then at each switching time t_i of $P(i = 1, 2, ..., k)$. The Hamiltonian H is indifferent to the choice of the control, which is:

$$H(\lambda(t_i), x(t_i), 1, t_i) = H(\lambda(t_i), x(t_i), 0, t_i) \tag{1.35}$$

The relationship between the "maximum principle" and the "indifference principle" is that the "indifference principle" can be implied by the "maximum principle". The "maximum principle" makes a stronger condition, that is, the control is forced to switch by the "maximum principle" when the phase space orbit crosses the indifference curve (1.37), while the control is allowed to change by the "indifference principle" at the same point. That means the control can stay the same in a region of phase space and also be optimal. It is not allowed to change the value of control v until reaching the indifference curve (1.37) again. It is workable even though the new theory requires more candidate optimal control paths. When a cost of a switching control exists, the "maximum principle" should be replaced by the "indifference principle", and optimal control will also exist. The existence of an optimal solution can be proved by Theorem 2 in Blatt's paper.

This research originated from Blatt's work. The goal of this book involves some novel computational algorithms for solving (1.28-1.32) based on the theorems in Blatt's paper. The work on the cost analysis, difference of optimal control policy sequences and division of the time intervals are extended from Blatt's original work. The methods, which could avoid the solution staying at local minimum without reaching the global minimum, are also discussed. This book is more concerned with using the computer software package to solve the problems rather than solutions analysis.

There are some computing optimal control methods (see next section) which have been successfully applied in many fields. They involve subdividing the interval $[0, T]$ into many (usually equal) subintervals. However, the accuracy will be lost if the switching times do not lie on the end-points of the equal subintervals. Hence it is essential to compute the optimal switching times.

7. Different Approaches to Optimal Control Problems

With the advances of modern computers and rapid development of software engineering, more and more people are concerned with computational algorithms, which could shorten computing time and provide more accurate results for complex optimal control problems that are difficult to solve analytically. During last thirty years, many efficient approaches have been developed and successfully applied in many models in a wide range of fields. Several numerical methods are available in the references ([24, 1981]; [56, 1975]; [57, 1986]; [58, 1986]; [76, 1981]; [75, 1980]; [84, 1991]; [85, 1991]), while some typical computing optimal control problems and efficient computational algorithms relevant to the present study will be discussed in the next few sections, the general computational approaches and algorithms for optimal control may be classified as the following (Islam [36, 2001]).

There is a wide range of algorithms which can be used for computing optimal control models in finance and can be classified under the algorithms for continuous and discrete optimal control models (Islam [36, 2001]). Algorithms for continuous optimal control models in finance include: (i) gradient search methods; (ii) algorithms based two value boundary problems; (iii) dynamic programming, approximate solution methods (steady-state solution, numerical methods based on approximation and perturbation, contraction mapping based algorithm and simulation); and (iv) control discretization approach based on step function, spline etc. Algorithms for discrete optimal control models in finance may be classified as follows: (i) algorithms based on linear and non-linear programming solution methods; (ii) algorithms based on the difference equations of the Pontryagin maximum principle and solved as a two value-

Optimal Control Models 11

boundary problem; (iii) gradient search method; (iv) approximation methods; and (v) dynamic programming.

For computing optimal control finance models, some recently developed computer packages such as SCOM, MATLAB, MATHEMATICA, DUAL, RIOTS, MISER and OCIM can be used.

7.1 OCIM

In reference [15, Craven 1998], a FORTRAN computer software package OCIM (Optimal Control in Melbourne) was discussed for solving a class of fixed-time optimal control problems. This computational method is based on the Augmented Lagrangian algorithm which was discussed in Section 6.3.2 in Craven [14, 1995]. Powell and Hestenes first made the Augmented Lagrangian for a problem with equality constraints. Rockafellar extended it to inequality constraints in paper [74, 1974]. OCIM can be run on a Macintosh as well as UNIX.

The basic idea of this method is to divide the time interval $[0, T]$ into n subintervals. The control $u(.)$ is approximated by a step-function as in MISER [33, 1987], with $u(t) = u_j$ (constant) on subinterval $[jT/n, (j+1)T/n]$. In MISER [33, 1987], Goh and Teo had obtained good numerical results using the apparently crude approximation of $u(.)$ by a step-function, which is called "control parameterization technique". In Craven [14, 1995], the theory in Section 7.6 and 7.8 proves that this occurs when the control system acts as a suitable low-pass filter; also the smoothing effect of integrating the state differential equation will often have this result. Increasing the number n of subdivisions will lead to a greater accuracy. Note in the calculation of the co-state equation for $\lambda(.)$, the interpolation of the values of state $x(.)$ is required. It is done by linear interpolation of the value of $x(.)$ at subdivision points $t = jT/n$. The linear interpolation is also used for calculating the gradient in this research.

A brief description of the computing method is discussed here. First, given the approximated control u to solve the state equation, obtain the state values taken at the grid-points of the subintervals; second solve the co-state equation with respect to $x(.)$; thirdly, calculate the augmented Lagrangian; and lastly, calculate the gradients of the cost function respect to $u(.)$. This algorithm had been programmed in FORTRAN. Note that jumps in co-states are not yet implemented. Unconstrained minimization is done using the CONMIN package [70], which uses the BFGS quasi-Newton algorithm [28, 1980].

Using the augmented Lagrangian allows gradients to be calculated by a single adjoint differential equation, and instead of having to code a separated formula for each constraint. In this algorithm, a time interval $[0, T]$ is divided into equal subintervals. A diversity of control problems can be solved by this method, but the equality constraint like $(\forall t)\gamma(x(t), u(t)) = 0$ presents some problems with OCIM. Several famous models were successfully solved by this method. They are a damped oscillator (the approaches developed in this research also solved an example of an oscillator problem), the Dodgem problem [29, 1975], and the Fish problem [12, 1976]. In Craven's paper, there was a useful non-linear transformation of the time scale that can effectively give a good subdivision in big time ranges, without requiring a large number of subdivisions, and also avoid computing problems. The details of the transformation will be described in Section 2.3.

7.2 RIOTS 95

Relating to another computing optimal control software package SCOM, which will be introduced in Section 1.7.5, the RIOTS (Recursive Integration Optimal Trajectory Solver 95) package [77, 1997] also runs on MATLAB. It solves optimal control problems by solving (in various ways) the differential equations that lead to function values and gradients, and then applying a minimizer package. The computing scheme is similar to the one that was described by Craven [14, 1995] in Section 6.4.5. This package obtains good accuracy of the switching times by using a sufficiently small subdivision of the time interval.

7.3 A computational approach for the cost of changing control

A computational method based on the control parameterization technique [84, 1991] was introduced in reference [82, 1992] for solving a class of optimal control problems where the cost function is the sum of integral cost, terminal cost, and full variation of a control. The full variation of a control is used to measure changes on the control action.

This method involves three stages of approximation. In the first stage, the problem is approximated by a sequence of approximation problems based on the control parameterization technique. Every approximation problem involving full variation of control is an optimal parameter selection problem. They become non-smooth functions in the form of l_1- norm. In the second stage, the non-smooth functions are smoothen by a smoothing technique [83, 1988]. Another smoothing technique [42, 1991] is used to approximate the continuous state inequality constraints by a sequence of conventional canonical constraints in stage three. All these approximation problems are standard optimal parameter selection problems which can be solved by the general optimal control software package MISER3 [40, 1991]. The method is supported by the convergence properties. A calculation of a fishery harvesting model (which was computed in reference [41, 1990] with penalty $c = 0$), was involved to illustrate the method, and the chosen value penalty term c was explained in this paper. A similar situation is also discussed in this book, which is called the chosen cost, in Section 2.8. An application of an optimal control problem in which the cost function does not have the full variation of control can also utilize this method by adjusting the penalty constant appropriately to obtain a smoother control without changing the optimum of the original cost function. There are some relevant references [55, 1987], [2, 1976], [66, 1977] for this problem.

7.4 Optimal switching control computational method

In Li [53, 1995], a computational method was developed for solving a class of optimal control problems involving the choice of a fixed number of switching time points $t_1, t_2, \ldots, t_K$ which divide the system's time horizon $[0, T]$ into K time periods. A subsystem S_i is selected for each time periods $[t_{i-1}, t_i]$, from a finite number of given candidate subsystems, which run in that time period. All the K subsystems and $K - 1$ switching times will be taken as decision variables. These decision variables form candidate selection systems , which will lead the optimal control problem to a minimum. Here, the optimal control problem is only considered over a finite time horizon. In this method, the original problem is transformed into an equivalent optimal control problem with system parameters by using some suitable constraints on the co-efficient of the linear combination, which is formed by the candidate subsystems, and using a time re-scaling technique.

Many control problems related to the system dynamics that are subject to sudden changes can be put into this model. In recent years, general optimal switching control problems have been studied. The optimality principle is applied and existence of optimal controls is discussed. Basically the problems are formulated as optimal control problems , in which the feasible controls are determined by appropriate switching functions. There are relevant references available in the bibliography [53, 1995], [87, 1989], [88, 1991], [27, 1979].

This situation has problems involving both discrete and continuous decisions represented by the subsystems and switching times. A transformed technique is introduced for solving this mixed discrete and continuous optimal control problem. The basic idea behind this technique is transforming the mixed continuous-discrete optimal control problem into an optimal parameter selection problem [84, 1991], which only deals with continuous decision variables. Since the transformed problem still involves the switching times located within subdivisions, which make the numerical solution of such an optimal control problem difficult. Another transformation is introduced to further transform the problem into an optimal control problem with system parameters. The control is taken as the lengths of the switching intervals as parameters. The second equivalent optimal control problem can be solved by standard optimal control techniques. A numerical problem was solved in Li's paper [53, 1995] by using this computational method.

This kind of optimal control problem has sudden changes in the dynamics at switching time, and therefore has a mixed continuous-discrete nature. Switching times, a time scaling and a control function are introduced to deal with the discontinuities in the system dynamics. The control function is a piecewise constant function with grid-points corresponding the discontinuities of the original problem, hence allowing the general optimal control software to solve the problem.

7.5 SCOM

In Craven and Islam [18, 2001] (See also Islam and Craven [38, 2002]), a class of optimal control problems in continuous time were solved by a computer software package called SCOM, also using the MATLAB system. As in the MISER [33, 1987] and OCIM [15, 1998] packages, the control is approximated by a step-function. Because of the matrix features in MATLAB, programming is made easier. Finite difference approximations for gradients give some advantages for computing gradients. In this paper, end-point conditions and implicit constraints in some economic models are simply handled.

Consider an optimal control problem of the form:

$$\text{MIN}_{x(.),u(.)} J^0(x,u) := \int_0^1 f(x(t),u(t),t)d + \Phi(x(1)) \qquad (1.36)$$

subject to:
$$x(0) = a, \dot{x}(t) = m(x(t),u(t),t) \ (0 \le t \le 1) \qquad (1.37)$$

$$g(u(t)) \le 0 \ (0 \le t \le 1) \qquad (1.38)$$

The end-point term and constraints can be handled by a penalty term; its detailed description will be introduced in Section 4.4. The concern here is only with how this computational algorithm works. The differential equation (1.39) with initial condition, determines $x(.)$ from $u(.)$; Denote $x(t) = Q(u)(.)$. The interval $[0, 1]$ is then divided into N equal subintervals, and $u(.)$ is approximated by a step-function taking values $u_1, u_2, \ldots, u_N$ on the successive subintervals. An extra constraint $u \in V$ is added to the given problem, where V is the subspace of such step-functions. Consequently, $x(.)$ becomes a polynomial function, determined by its values $x_0, x_1, \ldots, x_N$ at the grid-points $t = 0, 1/n, 2/N, \ldots, 1$. Since there are discontinuities at the grid-points on the right side of the differential equation which is a non-smooth function of

Optimal Control Models   15

t, a suitable differential equation solver must be chosen for such functions. Many standard solvers do not have this feature. Only one of the six ODE solvers in MATLAB is designed for stiff differential equations. However, this ODE solver is not used in this book. Instead, a good computer software package "nqq" is used to deal with the jumps on the right side of the differential equation in the dynamic system of the optimal control problems. The fourth order Runge-Kutta method [30, 1965] (a standard method for solving ordinary differential equations) is slightly modified for solving a differential equation of the form $\dot{x}(t) = m(x(t), u(t))$, where $u(.)$ is a step-function. The modification is simply recording the counting number j and time t to make sure that $u(t)$ always takes the appropriate value u_j not u_{j+1} in subinterval $[j/N, (j+1)/N]$ when $t = (j + 1)/N$. In the computation, the differential equation is computed forward (starting at $t = 0$), while the adjoint equation is solved backward (starting at $t = 1$).

Two steps are introduced to solve such optimal control problems:

1 Compute objective values from the objective function, differential equation, and augmented Lagrangian, not compute gradients from the adjoint equation and Hamiltonian. That assumes gradients can be estimated by finite differences from what to be computed.

2 Compute objective values and gradients from the adjoint equation and Hamiltonian.

Implicit Constraints: in some economic models, such as the model [48, 1971] to which Craven and Islam have applied the SCOM package in the paper [18, 2001], fractional powers of the functions (with $x(t)^\beta$ in the right side of the differential equation where $0 < \beta < 1$) appear, then some choices of $u(.)$ will lead to $x(.) < 0$, causing the solver to crash. The requirement of $x(.) \geq 0$ forms *an implicit constraint*. A finite-difference approximation to gradients is useful as approximations over a wider domain in this case. As mentioned in Section 1.7.1, linear interpolation can also be used in solving gradients and co-state functions. Increasing the number of the subintervals N will get better results. It will be discussed later in Section 2.7 and 2.8.

Two problems were tested using SCOM in a paper by Craven and Islam [18, 2001], and accurate results were found for both. In Craven and Islam [18, 2001], "constr" estimated gradients by finite differences to compute the optimum. The economic models in this paper [18, 2001] with implicit constraints are different from the models that were solved by other computer software packages. However, this computer software also needs to be further developed for more general use.

7.6 Switching Costs Model

An investment model for the natural resource industry was introduced in Richard and Mihall's paper [73, 2001] with switching cost. The problem combines both absolutely continuous and impulse stochastic control. In particular, the control strategy involves a sequence of interventions at discrete times. However, this component of the control strategy does not have all the features of impulse control because the sizes of the jumps associated with each interventions strategy are not part of the control strategy but are constrained to the pattern..., $1, -1, 1, -1,$..., jumping between two levels. This kind of control patterns is also considered in this research.

Perthame [68, 1984] first introduced the combination of impulse and absolutely continuous stochastic control problems which have been further studied by Brekke and B.Øksendal [3, 1991]. Mundaca and Øksendal [62, 1998] and Cadenillas and Zapatero [6, 2000] work on the applications to the control of currency exchange rates.

7.7 CPET

Time optimal control problems (which are not considered in this research) with bang-bang control associated with or without singular arc solutions can make the calculation difficult. A novel problem transformation called the Control Parameterization Enhancing Transform (CPET) was introduced in reference [51, 1997] to provide a computationally simple and numerically accurate solution without assuming that the optimal control is pure bang-bang control for time optimal control problems. A standard parameterization algorithm can calculate the exact switching times and singular control values of the original problem with CPET. Also, with the CPET, switching points for the control match the control parameterization knot points naturally, and hence piece-wise integration can be conveniently and accurately carried out in the usual control parameterization manner.

Several models used this technique and gave numerical results with extremely high accuracy. They are F-8 flight aircraft (first introduced in reference [31, 1977]) [71, 1994], the well-known "dodgem car" problem [29, 1975], and a stirred tank mixer [34, 1976]. The generalizations of CPET technique to a large range of problems were also introduced in reference [72, 1999].

7.8 STC

A new control method, the switching time computation (STC) method, which finds a suitable concatenation of constant-input arcs (or, equivalently, the places of switchings) that would take a given single-input non-linear system from a given initial point to the target, was first introduced in Kaya and Noakes' paper [43, 1994]. It was also described in detail of the mathematical

Optimal Control Models 17

reasoning in paper [44, 1996]. The method is applicable to single-input non-linear systems. It finds the switching times for a piecewise-constant input with a given number of switchings. It can also be used for solving the time-optimal bang-bang control problem. The TOBC algorithm, which is based on the STC method, is given for this purpose. Since the STC method is basically designed for a non-linear control system, the problem of the initial guess is equally difficult when it is applied to a linear system. For the optimization procedure, an improper guess for the arc times may cause the method to fail in linear systems. However, the initial guess can be improved by experience from the failures. In non-linear systems, there does not exist a scheme for 'guessing' a proper starting point in general optimization procedures. In general, the STC method handles a linear or a non-linear system without much discrimination. The reason is that the optimization is carried out in arc time space and even a linear system has a solution that is complicated in arc time space. The STC method has been applied as part of the TOBC algorithm to two ideal systems and a physical system (F-8 aircraft). They have been shown to be fast and accurate. The comparisons with results obtained through MISER3 software have demonstrated the efficiency of the STC method, both in its own right in finding bang-bang controls and in finding time-optimal bang-bang controls when incorporated in the TOBC algorithm. There is also a possibility to generalize the system from a single-input system to the multi-input system, which needs more computer programming involving.

7.9   Leap-frog Algorithm

Pontryagin's Maximum Principle gives the necessary conditions for optimality of the behavior of a control system, which requires the solution of a two-point boundary-value problem (TPBVP). Noakes [64, 1997] has developed a global algorithm for finding a geodesic joining two given points on a Riemannian manifold. A geodesic problem is a special type of TPBVP. The algorithm can be viewed as a solution method for a special type of TPBVP. It is simple to implement and works well in practice. The algorithm is called the *Leap-Frog Algorithm* because of the nature of its geometry. Application of the Leap-Frog Algorithm to optimal control was first announced in Kaya and Noakes [45, 1997]. This algorithm gave promising results when it was applied to find an optimal control for a class of systems with unbounded control. In Kaya and Noakes' paper [46, 1998], a direct and intuitive implementation of the algorithm for general non-linear systems with unbounded controls has been discussed. This work gave a more detailed and extended account of the announcement. A theoretical analysis of the Leap-Frog Algorithm for a class of optimal control problems with bounded controls in the plane was given in Kaya and Noakes' paper [47, 1998]. The Leap-Frog Algorithm assumes that the problem is already solved locally. This requirement translates to the case of optimal control as the availability of a local solution of the problem. This is related to the structure of the small-time reachable sets.

7.10 An obstruction optimal control problem

In Craven [17, 1999], an optimal control problem relating to flow around an obstacle (original proposed by Giannesi [32, 1996]) can be treated as a minimization problem which leads to a necessary condition or an optimal control problem. In this paper, Craven gave the discretization augment, and proved that, if an optimal path exists, the Pontryagin principle can be used to calculate the optimum. The optimum was verified to be reached by a discretization of the problem, and was also proved to be a global minimum.

7.11 Computational approaches to stochastic optimal control models in finance

Computational approaches specific to stochastic financial optimal control models are relatively well developed in the literature. However, computational approaches to deterministic financial optimal control models are not well documented in the literature, the standard general computational approaches to optimal control discussed above are applied to financial models as well. Some discussion of the computational approaches with specific applications to finance may be seen in Islam and Craven [38, 2002].

7.12 Comparisons of the methods

While a discussion of the comparisons of the general computational approaches is provided below, such comparisons are also relevant when the general approaches are applied to financial models. In financial optimal control models, the control function $u(t)$ is approximated by a vector on some vector space of finite dimension in all algorithms for numerical computation of such an optimal control model 1.38-1.40. There are some examples with different chosen approximations. The RIOTS 95 package [77, 1997] which uses MATLAB, uses various spline approximations, solves the optimization problems by projected descent methods; MISER3 [33, 1987], uses a step-function to approximate the control function, solves the optimization problems by sequential quadratic programming; OCIM [15, 1998], uses conjugate gradient methods. Different implementations behave differently especially on the functions defined on a restricted domain, since some optimization methods might want to search the area outside the domain. Although a step-function is obviously a crude approximation, it produces accurate results shown in many instants in reference [84, 1991]. Since integrating the dynamic equation $(d/dx)x(t) = \ldots)$ to obtain $x(t)$ is a smooth operation, the high-frequency oscillations are attenuated. In Craven [14, 1995], if this attenuation is sufficiently rapid, the result of step-function approximations converges to the exact optimum while $n \to \infty$. It is necessary to have some assumption of this qualitative kind in order to ensure that the chosen finite dimensional approximations will permit a good approximation to the exact optimum. The RIOTS 95 package can run faster than SCOM, maybe because of its implementation in the C programming language. The efficiency of the STC method has been demonstrated by comparisons with results through MISER3 optimal control software (a general-purpose optimal control software package incorporating sophisticated numerical algorithms). MISER3 did not get results as fast as the STC method did, perhaps because the general-purpose might be hampering its agility to certain extent because of some default settings regarding the tolerances for the accuracy of the ODE solver and optimization routine in the software.

This research is only concerned with pure bang-bang control problems within a fixed-time period. All the algorithms and transformations are made for this purpose. The control function is also approximated by a step-function. However, because the control does not always jump at the grid-points of the subdivisions of the time intervals which are usually equally divided in other works, it is necessary to calculate the optimal divisions of the time horizon. This research is mainly computing the optimal ranges of the subdivisions in time period as well as calculating the minimum of the objective function. Situations when a cost of changing control is involved in the cost function are discussed as well as how this cost can effectively work on the whole system. Although the STC method is also concerned with the calculation of the optimal switching times, it does not include the cost of each switching control.

The limitations of the above computational approaches are summarized in Chen and Craven [10, 2002]. From the above survey it will also appear that each of the above computational methods has characteristics which are computationally efficient for computing optimal control financial models with switching times. A new approach which can adapt various convenient components of the above computational approaches is developed in the next section. Although the present algorithm has similarity with CPET, the details of the two algorithms are different. A new computer package called CSTVA is also developed here which can suitably implement the proposed algorithm. The present computational method consisting the STV algorithm and the CSTVA computer programs does, therefore, provide a new computational approach for modeling optimal corporate financing. The computational approach can be suitably applied to any other disciplines as well.

This approach (STV) consists of several computational methods (described in Chapter 2):

1 The STV method where the switching time is made a control variable optimal value of which is to be determined by the model.

2 A piecewise-linear (or non-linear) transformation of time.

3 The step function approach to approximate the control variable.

4 Finite difference method for estimating gradients if gradients are not provided.

5 An optimization program based on the sequential quadratic programming (SQP) (as in MATLAB's "constr" program similar to the Newton Method for constrained optimization).

6 A second order differential equation to represent the dynamic model.

8. Conclusion

This book is mainly concerned with using the computational algorithms to solve certain classes of optimal control problems. The next chapter will introduce the computational approach named Switching Time Variable (STV) algorithm developed in this book that can solve a class of financial optimal control problems when the control is approximated by a step-function. The piecewise-linear transformation that is constructed for the computer software package is described in Section 2.2. Some non-linear transformations, which were first introduced by Craven [14, 1995], are also discussed in Section 2.3. These transformations are used to solve the large time period optimal control problem in Chapter 4. A computer software package that was developed by Craven and Islam [18, 2001] and Islam and Craven [37, 2001] is presented in Section 2.4. The "nqq" function for solving differential equations is quoted as a part of the computer software package in this research. The thrust of this book involves a general computer software for certain optimal control problems. The principal algorithms behind it are introduced in Section 2.6. All the computing results of an example problem for optimal investment planning for the economy are shown in graphs and tables in Section 2.7. A cost of changing control is also discussed in Section 2.8.

Chapter 2

THE STV APPROACH TO FINANCIAL OPTIMAL CONTROL MODELS

1. Introduction

In this chapter, a particular case of the optimal financial control problems, which has one state function and one control function that is approximated by a step-function, is discussed. Before the problem is defined, it is necessary to cover some concepts and transformations in Section 2.2 and Section 2.3, to explain the problems and algorithms that will be introduced in later sections and chapters. As part of the computer software package SCOM (that was constructed by Craven and Islam), "nqq" is used as a DE solver in the algorithms in this research. This program is described in Section 2.4. Then a simplified control problem is introduced in Section 2.5. The computational algorithms, which are used to solve this simplified control problem, are indicated in Section 2.6. Some problems with different fitting functions will be discussed later in Chapter 5. Lastly, graphs and tables in Section 2.7 represent all the computing results of this problem. The analysis of the results is discussed in Section 2.8.

2. Piecewise-linear Transformation

The idea of piecewise-linear transformation of the time variable was first introduced by Teo [40, 1991], but the time intervals were mapped into $(j, j+1)$ instead of $(jh, (j+1)h)$, $h = 1/n$, where n is the number of time intervals. In Lee, Teo, Rehbock and Jennings [51, 1997], the time transformation is described by $dt/ds = v(s)$, where v is a piece-wise constant transformation. In this book, a similar idea is used, but the implementation is a little simpler, not requiring another differential equation. A non-linear transformation of the time scale given by Craven [15, 1998], is introduced in Section 2.3. The transformations in this chapter are the essential part of the algorithms in the research, and will be directly applied in Section 2.6.

Consider a financial optimal control model, to minimize an integral:

$$\text{MIN } J = \int_0^1 [x(t) - \phi(t)]^2 dt \tag{2.1}$$

subject to:

$$\dot{x}(t) = u(t), \, x(0) = x_0 \tag{2.2}$$

$$0 \leq u(t) \leq 1, \, 0 \leq t \leq 1 \tag{2.3}$$

The time interval [0,1] is divided as follows:

$$0 = t_0 < t_1 < t_2 < \ldots < t_j < t_{j+1} < \ldots < t_{r-1} < t_r = 1;$$

so there are r subintervals:

$$[t_0, t_1], [t_1, t_2], \ldots, [t_{j-1}, t_j], \ldots, [t_{r-1}, t_r].$$

A scaled time τ is constructed here to replace the real time t in computation; τ will be used later in the "nqq" package (differential equation solver).

The scaled time τ takes values:

$$\tau = 0, h, 2h, \ldots, jh, (j+1)h, \ldots, (r-1)h, rh = 1$$

where $h = 1/r$, thus the time intervals [0, 1] is divided into r equal intervals.

The relationship between τ and t can be expressed by the following formula:

$$t = \varphi(\tau) = h^{-1} * (t_{j+1} - t_j) * (\tau - h * (j-1)) + t_j \tag{2.4}$$

where $\tau \in [0, 1]$.

Thus the subdivision point $t = t_j$ maps to point $\tau = jh$.

Now the control $u(t)$ takes values:

$$u(t) = u_0, u_1, u_2, \ldots, u_j, \ldots, u_{r-1}$$

in successive subintervals:

$$[0, h], [h, 2h], [2h, 3h], \ldots, [jh, (j+1)h], \ldots, [(r-1)h, 1]$$

Define $x(t) := x(\varphi(\tau))$.

Then the dynamic equation $x(0) = x_0$, $\dot{x}(t) = u(t)$ transforms to:

$$\begin{aligned} \dot{x}(t) &= \dot{x}(\varphi(\tau))(d/d\tau)\varphi(\tau) \\ &= u(t) := h^{-1} * (t_{j+1} - t_j) * u_j \\ &\quad (\text{for } jh < \tau < (j+1)h) \end{aligned} \quad (2.5)$$

The objective function transforms to the sum of integrals:

$$J = \sum_{j=0}^{r-1} J_j = \sum_{j=0}^{r-1} \int_{jh}^{(j+1)h} [x(\varphi(\tau)) - \phi(\varphi(\tau))]^2 h^{-1} * (t_{j+1} - t_j) d\tau \quad (2.6)$$

3. Non-linear Time Scale Transformation

In some time-optimal control problems or optimal control problems that are over a long time interval, the large number of subintervals of the time scale will cause computational difficulty. The optimal control functions lead to stable values monotonically when time t becomes large. So it would be useful if a suitable non-linear transformation of time scale is practical, (see reference [15, 1998]). With a suitable non-linear time scale, a few subintervals may be enough to get the same accuracy of large subintervals.

Goh and Teo [33, 1987] introduced a change of time scale for a time-optimal control problem while the terminal time T is a variable. Let $t = T\tau$ where τ is the new time variable mapping in [0,1]. t is again written for τ, then problem becomes a fixed-time optimal control problem with time interval [0, 1] and parameter T. This allows T to be computed accurately without being interpolated between subdivision points.

This makes the original financial optimization problem as the following:

$$\text{MIN}_{x(.),u(.)} \int_0^T f(x(t),u(t),t))dt + \Phi(x(T))$$

subject to:

$$x(0) = x_0, \dot{x}(t) = m(x(t),u(t),t)) \ (0 \leq t \leq T)$$

$$g(x(t),u(t)) \leq b(t) \ (0 \leq t \leq T)$$

$$q(x(T)) = 0$$

Transform into:

$$\text{MIN}_{x(.),u(.)} \int_0^1 f(x(t),u(t),t))Tdt + \Phi(x(1))$$

subject to:

$$x(0) = x_0, \dot{x}(t) = m(x(t),u(t),t))T \ (0 \leq t \leq 1)$$

$$g(x(t),u(t)) \leq b(t) \ (0 \leq t \leq 1)$$

$$q(x(1)) = 0$$

This transformation will be used later in the computational methods 4.1 - 4.5 for the financial model in Chapter 4.

Now consider, an objective function with a discount factor $e^{-\alpha t}$ where α is positive, thus:

$$\int_0^T e^{-\alpha t} f(x(t),u(t))dt$$

Define time t as:

$$t = -\alpha^{-1} log(1 - \beta \tau), \text{ where } \beta = 1 - e^{-\alpha T}, \tau \in [0,1]$$

The objective function becomes:

$$(\beta/\alpha) \int_0^1 f(x(t),u(t))dt$$

and the differential equation changes from $\dot{x}(t) = m(x(t), u(t), t)$ to:

$$\dot{x}(t) = \frac{(\beta/\alpha)m(x(t), u(t))}{(1 - \beta t)} \quad (0 \leq t \leq 1), \quad x(0) = x_0$$

For T is fixed and large, the denominator satisfies $1 - \beta\tau \geq 1 - \beta > 0$. If T is a variable to be optimized over, then $1 - \beta$ replaces the variable T, in an interval $0 < \epsilon \leq 1 - \beta \leq \frac{1}{2}$.

For example, in a fixed-time optimal control problem, if there is some "distinguished time" $\hat{T} \in [0, T]$, then an additional constraint $r(x(\hat{T})) = 0$ must be satisfied. A transformation of $t \in [0, T]$ to $\tau \in [0, \frac{1}{2}]$ maps $\hat{T}$ to $\frac{1}{2}$, thus to a subdivision point. In particular:

$$\tau = \frac{1}{2}t/\hat{T} \quad (0 \leq t \leq \hat{T}), \quad t = 2\hat{T}\tau \quad (0 \leq \tau \leq \frac{1}{2})$$

$$\tau = \frac{1}{2} + \frac{1}{2}\frac{(t-\hat{T})}{(T-QT)} \quad (\hat{T} \leq t \leq T), \quad t = \hat{T} + 2(T-\hat{T})(\tau - \frac{1}{2}) \quad (\frac{1}{2} \leq \tau \leq 1)$$

At $\tau = \frac{1}{2}$, the piecewise-linear transformation is not differentiable, and the co-state may be discontinuous.

4. A Computer Software Package Used in this Study

The computer package SCOM ([18, 2001]) is a tool to solve step-function optimal control problems on MATLAB 5.2, on a Macintosh computer. It is noted that a program written and compiled in C language will run faster than a MATLAB program for the same computation. However MATLAB is a matrix computation language, so it requires much less programming work in calculating matrix operations on MATLAB than other computer languages, such as C language and Fortran, etc. The program "constr" is a constrained optimization package in MATLAB's Toolbox, based on a SQP (sequential quadratic programming) method. "constr" will use gradients if supplied; otherwise it will estimate gradients by finite difference. In non-linear programming methods, the SQP method is very successful. The method closely mimics Newton's method for constrained optimization, just as it is done for unconstrained optimization. Using a quasi-Newton updating method, an approximation is made of the Hessian of the Lagrangian function at each major iteration. It is then used to generate a Quadratic Programming sub-problem whose solution is used to form a search direction for a line search procedure. In the research, we build another subroutine to be called as the objective function in "constr".

The state functions in this research were solved by the differential equation solver "nqq" in the SCOM package (see details in Appendix C). When results of the state come out on the grid-points, the objective function $J(u)$ becomes a function $\tilde{J}(u_1, u_2, \ldots, u_N)$ of N variables that can also be calculated by "nqq".

When the subroutines for the state functions and objective function are constructed, the MATLAB program "constr" is then used to obtain the optimal solution of the problem with respect to optimal switching times. In the program, the control u is approximated by a step-function.

5. An Optimal Control Problem When the Control can only Take the Value 0 or 1

From the point of view of control theory, the bang-bang optimal control happens when the systems of the optimal control problems are linear in control. The Nerlove-Arrow model [63, 1962] is an example of a bang-bang control following a singular arc control. Now, introduce a typical bang-bang optimal control problem.

We consider an optimal control problem:

$$\text{MIN} J(u) = \int_0^T [x(t) - \phi(t)]^2 dt \qquad (2.7)$$

subject to:
$$x(0) = 0 \qquad (2.8)$$
$$\dot{x}(t) = u(t) \qquad (2.9)$$
$$u(t) = 0 \text{ or } 1 \ (0 \leq t \leq 1) \qquad (2.10)$$

In this case, the state function $x(.)$ is assumed piece-wise smooth, and the control $u(t)$ jumps many times between 0 and 1 during the time interval $[0, T]$. Here, simplify the problem to $T = 1$. Now use a step-function:

$$v(t) = \frac{1}{2} + (-1)^k [v(0) - \frac{1}{2}] \text{ for } k = 0, 1, 2, \ldots, n$$

$$v(0) = 0 \text{ or } 1$$

to approximate the control $u(t)$. Here the target function is: $\phi(t)$. Let $\phi(t) = \frac{1}{2}t$.

Observe that if having instead $u(t) \in [0, 1]$, then $u(t) = \frac{1}{2}$ $(0 \leq t \leq 1)$ would be optimal, with $J(u) = 0$. But if $u(t)$ must be either 0 or 1, then

$u(t)$ will jump many times between 0 and 1. Suppose a further restriction is made that, for some given integer n, time interval $[0, 1]$ is divided into n equal subintervals $[0, 1/n], [1/n, 2/n], \ldots, [j/n, (j+1)/n], \ldots, [(n-1)/n, 1]$; ($u(t)$ takes a constant value (1 or 0) on each subinterval $[j/n, (j + 1)/n]$, ($j = 0, 1, \ldots, n − 1$). So if $u(t)$ takes the values in this pattern $1, 0, 1, 0, \ldots, 1, 0$ on the successive subinterval, then each subintervals contributes $1/(12n^3)$ to $J(u)$, $J(u) = 1/(6n^2) \to 0$ as $n \to 0$. But the optimal value $J = 0$ is not reached, unless an infinite number of jumps are allowed.

Now a term $K * n$ is added to the objective function. (2.7) is modified as follows:

$$\text{MIN } F = J(u) + kn = \int_0^T [x(t) - \phi(t)]^2 dt + Kn$$

Here, K is the cost of changing control and n is the number of time intervals. The objective function of the optimal control problem becomes a cost function. The algorithms in the next section are used for solving these kind of optimal control problems.

6. Approaches to Bang-Bang Optimal Control with a Cost of Changing Control

In this section, the computational methods for solving bang-bang optimal control problems with a cost of switching control are introduced. Simply modifying certain parts of these methods can satisfy a class of similar problems.

Suppose the number N of switching times is fixed by $N = nn$, say switching times $t_1, t_2, \ldots, t_{nn}$. Now consider (2.7) as a function of the switching times, say $J(u) = Q(t_1, t_2, \ldots, t_{nn})$. Then the minimum of this function is computed, starting from a small value of N. Here the components of the vector $um = (um_1, um_2, \ldots, um_{nn})$ are the lengths between each switching time. The vector um must satisfy the upper and lower bounds 0 and 1, thus $0 \le t_{j-1} \le t_j \le 1$, and a constraint $\sum_{i=1}^{N} um_i = 1$ must be satisfied. The vector $xm = (xm_1, xm_2, \ldots, xm_{nn})$ represents the values of the state function takes at each switching time. The algorithms are as follows:

Algorithm 2.1 Main Program (see project1_1.m in Appendix A.1)

Step 1. Initialization. Set the maximum number of function evaluations, par, which is the system parameter of the MATLAB "constr" function, and also another system parameter $par(13) = 1$ (1 represents the number of the equation constraint in the minimization problem), and a vector of the parameters which are used in the whole subroutines, par = [number of the state components, number of control components, nn= number of total subintervals], arbitrary starting lengths of switching time intervals

$um0 = (um0_1, um0_2, \ldots, um0_{nn})$, and vector ul = lower bound of um, vector uu = upper bound of um, thus $uu \leq um \leq ul$. Initialize the value of the state function $xinit$.

Step 2. Call the MATLAB "constr" function. In turn, "constr" calls the "Minimizing Program" to calculate the minimization of the calling program with respect to the optimal vector um.

Step 3. Input the optimal result um into "Minimizing Program" again to obtain the results of the objective function J_{nn} (the last value of the calculation) and the state vector xm, corresponding to the optimal um.

Step 4. Attach a cost K to nn. Add $K * nn$ to the objective function (2.7) and calculate $F = J_{nn} + K * nn$.

Step 5. Set a bigger nn; then go back to Step 2; EXIT when the result of cost function J stops decreasing.

Algorithm 2.2: Minimizing Program (see project1_2.m in Appendix A.1)

Step 1. Initialization. Input vectors um, par and initial state $xinit$. Set the initial state $xm(1) = xinit$, and $nx = par(1)$, the number of the state components, $nu = par(2)$, the number of the control components; initialize scaled time $t = 0$, subinterval counter $it = 1$, $hs = 1/nn$(length of each equal subinterval). Choose the "Input function for dynamic equation" as the right side of equation (2.9) to calculate the differential equation, input um.

Step 2. Call SCOM package function "nqq" with the stated "Input function for dynamic equation" to solve the differential equation (2.9) of the state function $x(t)$. Tabulate the solutions as the components of the vector $xm = [x(1), x(2), \ldots, x(nn)]$.

Step 3. Set the initial state $zz = 0$. Set initial scaled time $t = 0$, subinterval counter $it = 1$ again, choose the "input function for integration calculation" for SCOM function "nqq" for solving the integration of the objective function (2.7), input vector xm and um.

Step 4. Call SCOM function "nqq" with the stated "Input function for integration calculation" to calculate:

$$J(u(t)) = \sum_{j=0}^{nn-1} J_i$$

where:

$$J_i = \int_{jh}^{(j+1)h} [x(\varphi((\tau))) - \phi(\varphi(\tau))]^2 h^{-1}(t_{j+1} - t_j) d\tau$$

$$h = 1/nn$$

when:
$$jh < \tau < (j+1)h$$

by solving the differential equation:

$$w(0) = 0, \dot{w}(t) = [x\varphi((\tau)) - \phi(\varphi(\tau))]^2 h^{-1}(t_{j+1} - t_j)]$$

when:
$$jh < \tau < (j+1)h$$

Tabulate the results in $w(.)$ as the components of vector $jm = [j(1), \ldots, j(nn)]$.

Step 5. Take the last result of the vector jm as the value of the objective function, and calculate the constraint function of "Minimizing program", which is: $g(um) = \sum_{i=1}^{N} um_i - 1$.

Algorithm 2.3: Input function for dynamic equation (see project1_3.m in Appendix A.1)

Step 1. Initialization. Input scaled time t, subinterval counter it, the length of subintervals hs, and vector um, and set the number of subintervals $nn = 1/hs$.

Step 2. Set the control policy as vector $u = [u(1), u(2), \ldots, u(nn)]$ with alternating values 1 and 0 (as in 2.10) in successive subintervals.

Step 3. Construct the right side of the differential equation for the state function using the piecewise-linear transformation in (2.5).

Algorithm 2.4: Input function for integration calculation (see project1_4.m in Appendix A.1)

Step 1. Initialization. Input scaled time t, subinterval counter it, the length of subintervals hs, vector xm representing the values of the state function at each switching time, and also a new initial state z for integral, and the vector um. Set the number of subintervals $nn = 1/hs$.

Step 2. Use the linear interpolation to get an estimate "xmt" of the state, in a time t between grid-points $0, h, 2h, \ldots, nn*h$, where $h = 1/nn$.

Step 3. Add up components of the lengths of the switching time intervals in um to obtain the switching times "t_j" in (2.4).

Step 4. Construct the right side of the equation (2.4) to obtain the time variable t.

Step 5. Calculate the integral in (2.6) at scaled time t.

When the number of switching time N increases, the cost function decreases because of the better approximation. While calculating a minimization problem $KN + J_N$, the term KN increases with N increasing, and J_N decreases with bigger N. It can be found that the cost of changing control is very critical in the cost function. A proper chosen cost K will efficiently lead the cost function to the minimum. The analysis of the cost will be discussed in later sections.

7. An Investment Planning Model and Results

In this section, the computational results reported by a set of graphs will be introduced to verify the algorithms developed in Section 2.6. First we will introduce the fitting function. In this example, the fitting function is set to be $0.4 * t$. The state function $x(t)$ is used to approximate this fitting function. The formulae for this financial optimal control model is shown as follows:

$$\text{MIN } J = \int_0^1 |x(t) - 0.4 * t^2| dt \qquad (2.11)$$

subject to:
$$\dot{x}(t) = u(t) \qquad (2.12)$$
$$x_0 = 0 \qquad (2.13)$$
$$u(t) = 1, 0 \qquad (2.14)$$
$$0 \leq t \leq 1 \qquad (2.15)$$

where x(t) = stock price, u(t) = the proportion of total investment in stocks compared to other forms of financial investment.

Although the above model is an illustrative model, it can, however, represent an interesting financial decision making problem. The state equation represents the dynamics of the price of a stock. It is assumed that the change in the price of a stock is determined by the proportion of allocation of total funds for purchasing a stock. The objective of this control problem is to determine the value of u(t) (which only takes the value of 1 or 0) which can optimize the objective function so as to minimize the deviation of the state variable from its target value specified. Therefore this model (2.11 to 2.15) is an investment planning model with some sub-utilization criterion included in the model.

The target function is $0.4*t^2$. First set $n = 2$ (n is the number of time intervals), and control takes as $1, 0$ in time intervals $[0, t_1], [t_1, 1]$. The model 2.11 to 2.15 was solved with these parameter values and the results are shown in Appendix B.1 and are represented by Figure 2.1. In Figure 2.1, it is shown that during the planning horizon $[0, 1]$, the control only jumps once at time $t = 0.127$. The state vector xm takes the value $[0.127, 0.127]$ at the grid-points of two subintervals $[0, 0.127]$ and $[0.127, 1]$. The approximation between the state function xm and the fitting function $0.4 * t^2$ is not very good when the number of the switching times is quite small (n is only 2). "$*-$" represents the state function $x(t)$ and "." represents the fitting function $0.4 * t^2$ in the following graph.

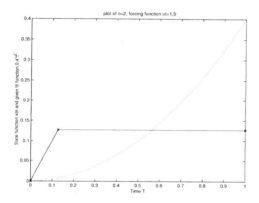

Figure 2.1. Plot of n=2, forcing function ut=1,0

Then increase n to 4. During the whole time [0,1], the control jumps three times. A better approximation is shown in Figure 2.2 with more jumps of the control.

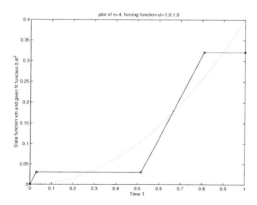

Figure 2.2. Plot of n=4, forcing function ut=1,0,1,0

Set $n = 6$, and run the program. A much better approximation than $n = 4$ is shown in Figure 2.3 because of the increased switching times.

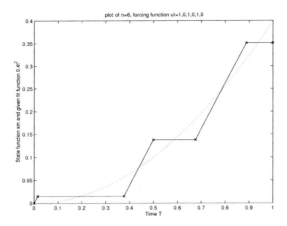

Figure 2.3. Plot of n=6, forcing function ut=1,0,1,0,1,0

When $n = 8$, a very close fit between state xm and the given fitting function $\phi(t)$ is shown in Figure 2.4. The result proves a very good convergence of the algorithms.

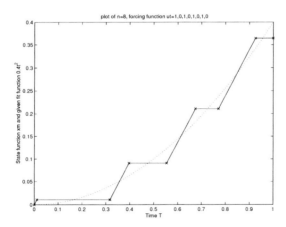

Figure 2.4. Plot of n=8, forcing function ut=1,0,1,0,1,0,1,0

In Figure 2.5, although the approximation between xm and $\phi(t)$ is still getting better when $n = 10$, the difference between the result of $n = 8$ and $n = 10$ is not as big as between $n = 2$ and $n = 4$. The decreasing of the results of the objective function slows down when n is very large.

The STV Approach to Financial Optimal Control Models 33

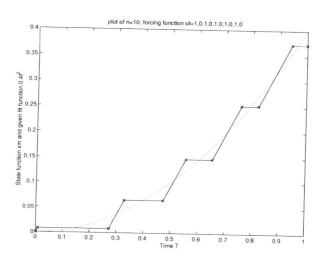

Figure 2.5. Plot of n=10, forcing function ut=1,0,1,0,1,0,1,0,1,0

The results of the objective function according to different numbers of the time intervals n are put in Table 2.1. The decreasing of the results of the objective function follows the increases number of the time intervals n. When n is small, the result of the objective function will decrease very fast, however this decrease will slow down when n becomes big. A good illustration is shown in Figure 2.6.

Table 2.1. Objective functions with the number of the switching times

n	J
2	0.062696
4	0.029085
6	0.018364
8	0.013369
10	0.010463

In Figure 2.6, the results of the objective function against the different number of the time intervals n are shown. From the graph, we will find out that the descent of the results of the objective function slows down when the number of time intervals n increases. The conclusion can be made that more jumps of the control in the time period $[0, 1]$ give a better association between the state $x(t)$ and fitting function $0.4t^2$ (i.e. makes the financial system more stable along its desired path) and leads the financial system to reach the best fit when n becomes infinity.

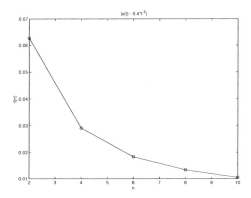

Figure 2.6. Plot of the values of the objective function to the number of the switching times

The optimal control problem in (2.11)-(2.15) is computed and the results are shown in the above graphs. The number of time intervals $n = 2, 4, 6, 8, 10$ is set consequently. A better fit comes into being Figures 2.1, 2.2, 2.3, 2.4 and 2.5. In Figure 2.6, it is shown that the result of the objective function J decreases while the number of time intervals n increases. Same results are also shown in Table 2.1. Now a proper cost K is set and attached with the number of the time intervals n to the objective function J. The objective function (2.11) becomes:

$$\text{MIN } F = J(n) + kn = \int_0^1 |x(t) - 0.4 * t^2| dt + Kn \quad (2.16)$$

Use the algorithms to solve this new optimal control problem with the same constraints as in (2.12)-(2.15). Three different values of cost K are chosen for $F(n) = J(n) + K * n$. The results of adding $K * n$ to the objective function are shown in Table 2.2, corresponding to $n = 2, 4, 6, 8, 12$.

Table 2.2. Costs of the switching control attached to the objective function

K	$n = 2$	$n = 4$	$n = 6$	$n = 8$	$n = 10$
K=0.02	0.102692	0.109085	0.138364	0.173369	0.210463
K=0.002	0.066696	0.037085	0.030364	0.029369	0.030463
K=0.0002	0.063096	0.029885	0.019564	0.014969	0.012463

From the results in Table 2.2, a conclusion can be made that only when cost K takes a certain value, it will lead the cost function to a minimum infinite switching times. When $K = 0.02$, the results of the cost function are increasing from $n = 2$ to $n = 10$, and the increases become faster. When $K = 0.0002$, the results are decreasing, but this decrease begins to slow down when n increases. It is conjectured that when the number of the switching times n is getting very big, this decreases will stop at a certain value of n. Meaningful results from $K = 0.002$ are given, and a minimum is obtained at $n = 8$. The following Figure 2.7 shows the results of the cost function against the number of the time intervals n while the cost $K = 0.002$ is attached to the objective function. The bottom of the line in the figure is the minimum point $n = 8$.

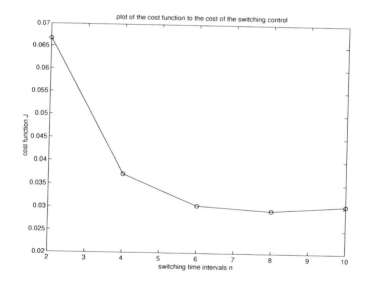

Figure 2.7. Plot of the cost function to the cost of switching control

This is an effective example of computational algorithms. All the results confirm the accuracy of computational algorithms in Section 2.6. A hand calculation of the part of the differential equation to exam the algorithms also gives a proof of computational accuracy. Some other experiments with different fitting functions will be discussed later in Chapter 5.

8. Financial Implications and Conclusion

The crucial aspect of the computational experiments undertaken here is that switching times and and the cost of switching times have significant implications for optimal investment planning. As the graphical results showed in Section 2.7, different numbers of switching times lead to different results for the objective function of the financial model. While the number of switching times increases, the value of the objective function decreases. That means a better fit is obtained. When a cost of changing control at each switching time is added to the original objective, the perspective changes. In the present case, when the number of switching times increases, the value of the term, which includes cost and the number of the switching, also increases. This increase slows down the decreasing of the original objective function. When the number of switching times increase to a certain value, the cost function stops decreasing and instead increases. The intermediate point between the decreasing and increasing is the optimal point.

The value of the cost of switching significantly influence the values of the switching times and optimal control. Comparing with the result of the objective function, if the cost is too big, the cost functional will increase all the time. It can be explained that it costs too much to change a control at each time interval. The financial system can never reach an optimal solution. But if the cost is too small, it could not affect the cost function at all. Then the control will jump infinitely in the time period to search an optimal solution, which is difficult to realize in real computation. Therefore, the value of the cost of switching time affects the optimal number of switching time. In terms of the value of the optimal control, it is found that several jumps in the strategy of investment in the stock market, shown by 1,0 values of investment in the stock market are optimal. A higher cost for switching reduces the optimal value of switching times compared to a situation when there is no cost for control switching.

An optimal investment strategy, therefore, should always be made on the consideration of the cost of control switching to determine how often the investment strategy can be changed between 1 and 0.

In the next chapter, a financial oscillator model for optimal aggregative investment planning will be presented. Since the oscillator problem has a state function, which is a second-order differential equation, the computational algorithms require transforming the second-order differential equation into two equivalent first-order differential equations. A new time scaled transformation for better integral calculation will be introduced in Section 3.3. The corresponding transformations of the state functions and the objective function will also be contained in Section 3.3. The computational algorithms for the financial oscillator problem will be described in Section 3.4. The computer software packages for these algorithms will be introduced in Appendix A.2. Different patterns of the control might lead the computation to different minimum points. The control with different initial starts was put into the program. The final result of the comparison will be the optimum of the control problem. The computing results of a particular oscillator problem are represented in graphs and tables in Section 3.6.

Chapter 3

A FINANCIAL OSCILLATOR MODEL

1. Introduction

The financial sector is very volatile, more volatile than the business cycle instabilities of the whole economy. Modeling the oscillatory dynamics of the financial sector is well developed. This type of dynamic financial system can also be modeled as an oscillatory optimal control model with dampening where a control law is specified in the model to make the financial system stable (see Sengupta and Fanchon [80, 1997]). The objective of this type of model involves the minimization of deviations of the system variables from their desired path. Such an oscillatory financial model is developed and computed in the chapter. These type of oscillatory financial models are useful in financial decision making since general optimal control models may not provide stabilization policies and the damped oscillator models provide an effective stabilization mechanism.

In Section 2.5, a financial optimal control model, which has one state function, one control function, and the control taking two constant values sequentially, was discussed. The computational algorithms were also developed and applied to this problem (For a general discussion of this approach see also Chen and Craven [10, 2002], as most of the materials in this chapter are adapted in this paper. However, while Chen and Craven [10, 2002] provide a discussion of the approach, the present chapter shows the application of this approach to financial modeling only.) Substantial results were obtained to verify the algorithms. In this chapter a more complicated optimal control problem is introduced in Section 3.2 that can represent the oscillatory behavior of the financial sector. The financial system includes a second order differential equation. A transformation of time scale used to obtain the optimal switching times as well as a better gradient is described in Section 3.3. The transformations of the control, state, and the objective function are changed corresponding to the new time subdivision. The computational methods for these kinds of optimal control problems are constructed in Section 3.4. In this chapter, the control is also approximated by a step-function. Additional analysis of the control with different patterns leading to different minimums is discussed in Section 3.5. In Section 3.6, two sets of graphical results with different control patterns are presented.

In economics problems, the dynamic behavior may require a second order differential equation and more than one control. In Blatt [2, 1976], some generalizations of multi-state functions and controls were discussed. But Blatt did not give any explicit proof and explanation of these generalizations. The computational algorithms in this chapter effectively solve the second order differential equation. This technique is also applied to an optimal corporate financing model later in Chapter 4.

2. Controlling a Damped Oscillator in a Financial Model

We consider an aggregate dynamic financial system described by a second-order differential equation as follows:

$$\ddot{x}_1(t) + \beta \dot{x}_1(t) + T^2 x_1(t) = T^2 u(t) \quad (3.1)$$

This financial system second order differential equation can be transformed into an equivalent pair of first-order differential equations:

$$\dot{x}_1(t) = T x_2(t)$$

$$\dot{x}_2 = T(-x_1(t) + u(t) - T x_2(t))$$

We now consider an oscillator problem with the bang-bang optimal control solution. A damped oscillator financial model for the above aggregate financial system with forcing function $u(.)$, and parameter T, is presented as follows (see another version of this financial model in Section 5.3.1):
/indexforcing function

$$\text{MIN } J = \int_0^T |x_1(t) - \phi(t)| dt \quad (3.2)$$

subject to:

$$\dot{x}_1(t) = T x_2(t) \quad (3.3)$$

$$\dot{x}_2 = T(-x_1(t) + u(t) - T x_2(t)) \quad (3.4)$$

A Financial Oscillator Model

$$x_1(0) = x_{10} \tag{3.5}$$
$$x_2(0) = x_{20} \tag{3.6}$$

where $x_1(t)$ = stock prices, $u(t)$ = the proportion of investment in stocks compared to other form of financial investment.

Here $\phi(.)$ is a given target function showing the desired path to be achieved by the financial system. Suppose that $u(.)$ is restricted by:

$$(\forall t)|u(t)| \leq 2 \tag{3.7}$$

Since the dynamics are linear in $u(.)$, if an optimum is reached, then bang-bang control may be expected, with possibly a singular arc for a time interval. However, singular arc control is not considered here.

The model (3.2) to (3.7) represents a financial decision making problem where the cost of changing control is added to the objective function, shown as follows:

$$\text{MIN } F = \int_0^T |x_1(t) - \phi(t)| dt + Kn$$

Here, K is the cost of switching control, n is the number of control jumps. The modified model calculates the optimal allocation for stocks with an explicit consideration of changing allocation shares. The cost function becomes the new objective function to be computed.

The model (3.2) to (3.7) is similar to the model (2.11) to (2.15) and represent a financial decision making problem of choosing the optimal allocation of finance in an economy for stocks. The forcing function is an additional feature of this model compared to the model in (2.11) to (2.15). To make the model realistic, it is necessary to incorporate the cost of changing control in this model.

3. Oscillator Transformation of the Financial Model

In this section, a good transformation of time scale is introduced. The computer package "nqq" is also used for solving the differential equations in this chapter. However, the "nqq" package does a limited job, it only does one Runge-Kutta step for each subinterval given to it. So when smaller subintervals are needed for accuracy, they must be supplied by the "small subintervals" construction. Although "nqq" package has this limitation, it has a great benefit for solving the differential equation with a non-smooth right-hand side of the differential equation which means there can be jumps at the end-points of intervals. Since a time-optimal control problem is not considered in this research, T is a constant. To make it easy, the time period of the optimal control problem described in last section is simplified to be $[0, 1]$ (the transformation of time scales when T is variable was given in Section 2.3).

First, the time horizon $[0, 1]$ is divided into nb "big subintervals", with end-points:

$$0 = pt_0 < pt_1 < \ldots < pt_j < pt_{j+1} < \ldots < pt_{nb-1} < pt_{nb} = 1$$

Since the functions change quite rapidly within each big subinterval, further subdivisions into "small intervals" are needed to get sufficient precision in solving the differential equations. Then each subinterval $[pt_j, pt_{j+1}]$ ($1 \leq j \leq nb$) is subdivided into ns "small subintervals" with end-points:

$$pt_j, pt_{j+\frac{1}{ns}}, pt_{j+\frac{2}{ns}}, \ldots, pt_{j+\frac{ns-1}{ns}}, pt_{j+1}$$

The whole subdivision of the time interval $[0, 1]$ is shown as follows:

$$0 = t_0 < t_1 < t_2 < \ldots < t_i < t_{i+1} < \ldots < t_{(nb*ns)-1} < t_{nb*ns} = 1$$

The relationship between pt and t can be represented as follows:

$$\forall j \in [0, nb] \sum_{i=j}^{j+ns} t_{i+1} - t_i = pt_{j+1} - pt_j$$

The control only jumps from one big interval to the next one. It takes values as follows:

$$\begin{aligned} u(t) = &\ u_0, u_0, \ldots, u_0, u_1, u_1, \ldots, u_1, u_2, u_2, \ldots, u_2, u_3, \ldots, \\ &\ u_{nb-2}, u_{nb-1}, u_{nb-1}, \ldots, u_{nb-1} \end{aligned} \quad (3.8)$$

A scaled time τ is constructed for the computer package, corresponding to the total number of subintervals $nn = nb * ns$, which takes values:

A Financial Oscillator Model

$$\tau = 0, h, 2h, \ldots, ih, (i+1)h, \ldots, (r-1)h, rh = 1$$

Here, $h = 1/nn, r = nn, nn = nb * ns$.

Two vectors which represent two different levels of the subdivision of the time period $[0, 1]$ um and sm are defined for mapping the control and calculating the integral (see the next section).

The relationship between the real time t which is used in the calculation of the integration and the scaled time τ is shown as follows:

$$t = \varphi(\tau) = h^{-1} * (t_{i+1} - t_i) * (\tau - h * (i-1)) + t_i,$$
$$\text{for } ih < \tau < (i+1)h, t_i < t < t_{i+1} \qquad (3.9)$$

where $\tau \in [0, 1]$, time $t \in [0, 1]$, and $h = 1/nn, nn = nb * ns$. Here, the control only jumps at the end-points of the big subintervals. Thus the control has the same constant value in all the small subintervals which make up a big subinterval. Another real time pt which represents the real switching times for the control is defined as follows:

$$pt = \hat{\varphi}(\tau) = nb^{-1} * (pt_{j+1} - pt_j) * (\tau - nb * (j-1)) + pt_j,$$
$$\text{for } ih < \tau < (i+1)h, pt_j < t < pt_{j+1} \qquad (3.10)$$

where, nb is the number of big subintervals. This pt is used in differential equation solving with respect to control $u(t)$. Correspondingly the control and state with respect to τ are defined:

$$u(t) := u(\hat{\varphi}(\tau)) \qquad (3.11)$$
$$x(t) := x(\hat{\varphi}(\tau)) \qquad (3.12)$$

The two first-order differential equations of the state functions of the financial system in (3.2) and (3.3) are transformed into two equivalent first-order equations:

$$\dot{x}_1(t) = \dot{x}_1(\hat{\varphi}(\tau))(d/d\tau)\hat{\varphi}(\tau) = T * x_2(\tau) * nb * (pt_{j+1} - pt_j) \qquad (3.13)$$

$$\begin{aligned}\dot{x}_2(t) &= \dot{x}_2(\hat{\varphi}(\tau))(d/d\tau)\hat{\varphi}(\tau)\\ &= T*(-x_1(\tau)+u_j-T*x_2(\tau))*nb*(pt_{j+1}-pt_j)\end{aligned} \quad (3.14)$$

The objective function of the financial optimal control model is then transformed to a sum of integrals in successive subintervals:

$$J = \sum_{i=0}^{r-1} J_i = \sum_{i=0}^{r-1} \int_{ih}^{(i+1)h} nn * |x_1(\hat{\varphi}((\tau))) - \phi(\varphi(\tau))| * (t_{i+1}-t_i) dt \quad (3.15)$$

where h = 1/nn, nn = nb*ns.

4. Computational Algorithm: The Steps

In this approach, time intervals of the financial model are defined at two levels, with the planning period $[0, T]$ first divided into nb big intervals, represented by vector um (see also Chen and Craven [10, 2002]):

$$um = um_1, um_2, \ldots, um_{nb-1}, um_{nb},$$

then each big subinterval is subdivided into ns small subintervals. Thus $[0, T]$ is subdivided into $nb * ns$ numbers of subintervals, represented by the vector sm:

$$sm = sm_1, sm_2, \ldots, sm_{ns}, sm_{ns+1}, \ldots, sm_{(nb-1)*ns+1},$$

$$sm_{(nb-1)*ns+2}, \ldots, sm_{nb*ns}$$

At each end-point of the big subintervals, the control jumps.

Algorithm 3.1 Main Program for the oscillator problem (see project2_1.m in Appendix A.2)

Step 1. Initialization. Set a vector of parameters par = [the number of the state components, number of control components, nb = the number of big subintervals, ns = the number of small subintervals]; and get the total number of the subintervals by calculating $nn = nb * ns$. Set the MATLAB "constr" function's system parameters $par(13) = 1$ (here, 1 represents only one equation constraint in the minimization problem), $par(14)$ = the maximum number of function evaluations; arbitrary starting lengths of the switching time intervals $um0 = (um0_1, um0_2, \ldots, um0_{nb})$. Set the vectors of upper

A Financial Oscillator Model

bounds uu and the vector of lower bounds ul of um, thus $uu \leq um \leq ul$. Also set the initial state $xinit$.

Step 2. Call the MATLAB "constr" function. In turn, "constr" calls the "Minimization Program" to calculate the minimization of the calling program with respect to the optimal vector um.

Step 3. Input the optimal result um into "Minimization Program", to obtain the values of the objective function $J_{(nn)}$ (the last value of the integral) and state vector xm (xm is a vector of all the values of the state functions take at the grid-points of the switching time intervals).

Step 4. Attach a cost K to nn. Add $K*nn$ to the objective function (3.2) and calculate $F = J_{nn} + K*nn$.

Step 5. Set a bigger nn and new starting time intervals, go back to Step 2; EXIT when the result of cost function J stops decreasing.

Algorithm 3.2: Minimization Program (see project2_2.m in Appendix A.2)

Step 1. Initialization. Input um, par and initial state $xinit$. Set the initial state $xm(1,:) = xinit$, and $nx = par(1)$, the number of the state components, $nu = par(2)$, the number of the control components, $nb = par(3)$, the number of big time intervals, $ns = par(4)$, the number of small time intervals, $nn = par(3)*par(4)$, the number of total time intervals; initialize the scaled time $t = 0$, subinterval counter $it = 1$, hs = length of each equation subintervals. Choose the "Input function for second order differential equation" as the right side of the differential equation (3.1), input um (in this chapter, the second order equation in the optimal control problem is transformed into two equivalent first order differential equations; so there are two state functions).

Step 2. Construct the vector sm whose components represent the lengths of the all time intervals by dividing each big time interval um by ns.

Step 3. Call the SCOM package function "nqq" with the stated "Input function for second order differential equation" to solve the two equivalent first order differential equations (3.3), (3.4) for the state function $x(t)$. Tabulate the solution for the state as the vector $xm(:,1) = [x_1(1), x_1(2), \ldots, x_1(nn)]$, $xm(:,2) = [x_2(1), x_2(2), \ldots, x_2(nn)]$, $xm(:,1)$ is the result of (3.3), $x(:,2)$ is the result of (3.4).

Step 4. Set the initial scaled time $t = 0$, subinterval counter $it = 1$, initial state $zz(:1) = 0$, $ma = 1$, the number of the input state function. Choose the "input function for the oscillator problem" for SCOM function "nqq" to solve the integration of the objective function (3.2). Input the vectors $xm(:1)$ and sm.

Step 5. Call SCOM function "nqq" with the stated "Input function for the oscillator problem" to solve the differential equation $d(J(w(t)))/dt = \dots$. Tabulate results in $w(.)$ as the components of vector $jm = [j(1),\dots,j(nn)]$.

Step 6. Take the last result of the vector jm as the value of the objective function, and calculate the constraint function of "Minimization program" which is $g(um) = \sum_{i=1}^{N} um_i - 1$.

Algorithm 3.3: Input function for second order differential equation (see project2_3.m in Appendix A.2)

Step 1. Initialization. Input scaled time t, subinterval counter it, the length of total subintervals hs and vector um. Set the value to parameters T, (set B for(3.18) and (3.19) in the next section). Set the number of the total subintervals $nn = 1/hs$.

Step 2. Obtain the number of the big time intervals by $nb = nn/ns$ for establishing the control policy.

Step 3. Set the control policy as vector $u = [u(1), u(2), \dots, u(nb)]$ with alternating value 1 and -1, see (3.8); the control only jumps at the end points of the big time intervals.

Step 4. Construct the right side of the transformation equation for time pt in (3.10): $pt = \hat{\varphi}(\tau) = nb^{-1} * (pt_{j+1} - pt_j) * (\tau - nb*(j-1)) + pt_j$, for $ih < \tau < (i+1)h$, $pt_j < t < pt_{j+1}$.

Step 5. Obtain the right sides of the differential equations by using the transformations in (3.13) and (3.14): $\dot{x}_1(t) = \dot{x}_1(\hat{\varphi}(\tau))(d/d\tau)\hat{\varphi}(\tau) = T * x_2(\tau) * nb * (pt_{j+1} - pt_j)$, $\dot{x}_2(t) = \dot{x}_2(\hat{\varphi}(\tau))(d/d\tau)\hat{\varphi}(\tau) = T*(-x_1(\tau) + u_j - T*x_2(\tau)) * nb * (pt_{j+1} - pt_j)$.

Algorithm 3.4: Input function for the oscillator problem (see project2_4.m in Appendix A.2)

Step 1. Initialization. Input scaled time t, subinterval counter it, the length of total subintervals hs, vector xm of values of first state function at switching times, and also the initial state z and vector sm. Set the number of the total subintervals $nn = 1/hs$.

Step 2. Use linear interpolation to get the estimate "xmt" of the state, in a time t between grid points $0h, h, 2h, \dots, h*nn$, where $h = 1/nn$.

Step 3. Add up the time intervals sm to get time "t_j" in (3.9): $t = \varphi(\tau) = h^{-1} * (t_{i+1} - t_i) * (\tau - h*(i-1)) + t_i$, for $ih < \tau < (i+1)h$, $t_i < t < t_{i+1}$.

A Financial Oscillator Model

Step 4. Construct the right side of the equation (3.9) to obtain time variable "t".

Step 5. Calculate the integrand in (3.15) at scaled time t: $J = \sum_{i=0}^{r-1} J_i = \sum_{i=0}^{r-1} \int_{ih}^{(i+1)h} nn * |x_1(\hat{\varphi}((\tau))) - \phi(\varphi(\tau))| * (t_{i+1} - t_i)dt$.

5. Financial Control Pattern

The financial dynamic system introduced in Section 3.2 is a second-order differential equation, and the forcing function /indexforcing function $u(t)$ takes values such as $-2, 2, -2, 2, \ldots$, switching between them at optimal switching times determined by an optimization calculation. Note if $u(t)$ is restricted to choose two values -2 and 2, then both the patterns $-2, 2, -2, 2, \ldots$ and $2, -2, 2, -2, \ldots$ have to be computed. The comparison of the solutions of these two patterns will be discussed in next section. This research studies a set of bang-bang controls that jump on the extreme points of the feasible area. However, since the sequences of the control switching are not defined, all the sequences of the control patterns need to be computed and analyzed. A numerical example in financial modeling is shown next.

6. Computing the Financial Model: Results and Analysis

In this section, the control $u(t)$ is set to be -2 and 2. The solutions of two different initial starts of control $u(t)$ are discussed and compared.

We can formulate the financial optimal control problem as follows:

$$MINJ = \int_0^1 |x_1(t) + 5t - 5|dt \qquad (3.16)$$

subject to:

$$\dot{x}_1(t) = T * x_2(t) \qquad (3.17)$$

$$\dot{x}_2(t) = -T * x_1(t) + T * u(t) - T^2 * B * x_2(t) \qquad (3.18)$$

$u(t)$ takes value $-2, 2, \ldots$ or $2, -2, \ldots$ in successive time intervals (3.19)

$$x_1(0) = 3 \qquad (3.20)$$

$$x_2(0) = 5 \qquad (3.21)$$

$$T = 5, B = 0.2 \qquad (3.22)$$

The target value of the target = $5t - 5$.

48   OPTIMAL CONTROL MODELS IN FINANCE

Table 3.1.   Results of the objective function at control pattern -2,2, ...

F	$n=2$	$n=4$	$n=6$	$n=8$
ns=1	1.1694	0.4538	0.3424	0.3069
ns=2	0.5767	0.2835	0.2494	0.2192
ns=4	0.4954	0.2311	0.1931	0.2054
ns=6	0.4581	0.2096	0.1832	0.1644
ns=8	0.4519	0.2492	0.2531	0.1701
ns=10	0.4437	0.2001	0.1681	0.1613
ns=16	0.4302	0.1896	0.1591	0.1541
ns=32	0.4187	0.1842	0.1514	0.1842
ns=64	0.4156	0.1828	0.1905	0.1459

Computational results are indicated in Table 3.1 above.

All the graphs shown below and on the following few pages are the computing results of the financial optimal control problem (3.16)-(3.22). There are three sets of the graphs, which represent the solutions at $nb = 2$, $nb = 4$, and $nb = 6$, where nb is the number of the big time intervals. In each set, the big time interval is subdivided by $ns = 1, 2, 4, 6, 8, 10$ respectively, where ns is the number of the small time intervals. Each set has six pairs of graphs representing six different subdivisions. All the graphs are shown in the same order. A pair of graphs represent "the state function $x_1(t)$ and the given fitting function against time t" and the comparison of two state functions, $x_1(t)$ and $x_2(t)$.

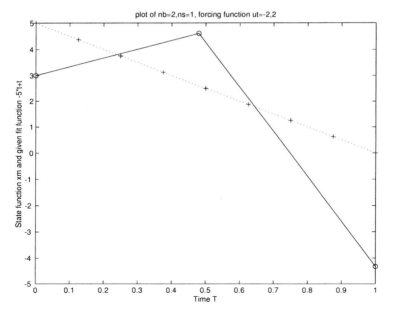

A Financial Oscillator Model

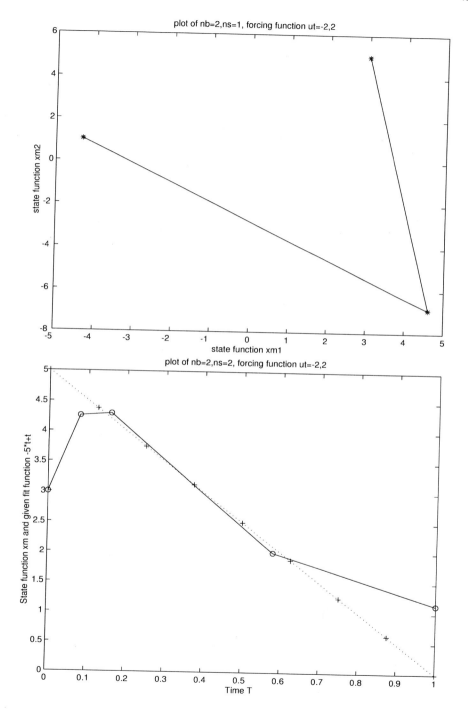

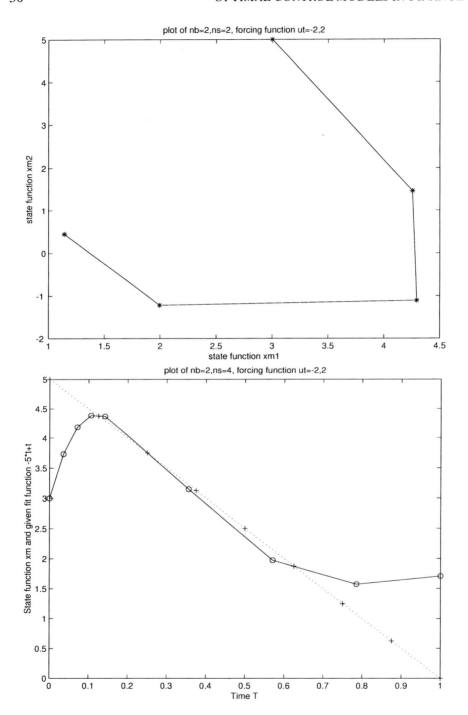

A Financial Oscillator Model

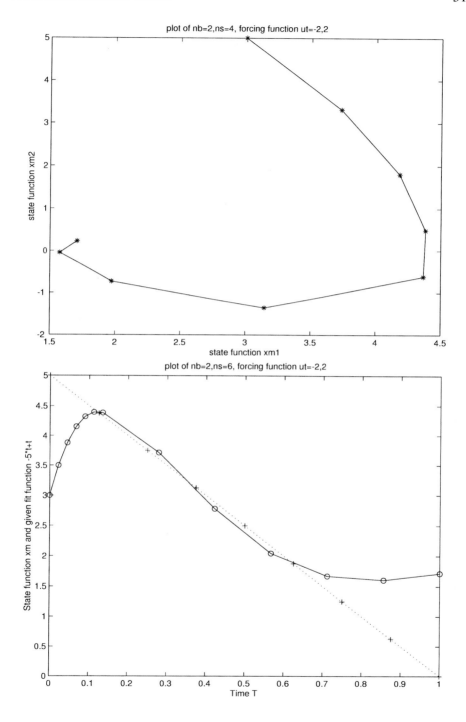

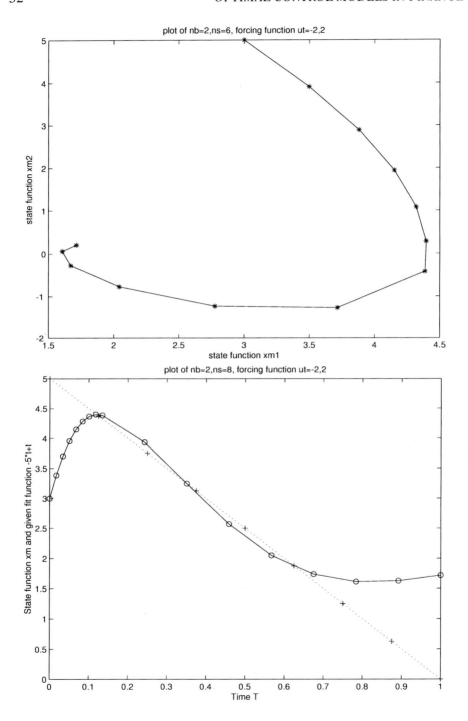

A Financial Oscillator Model

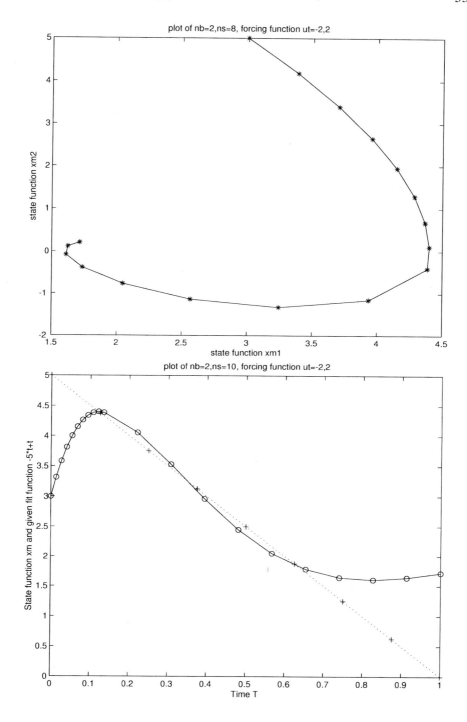

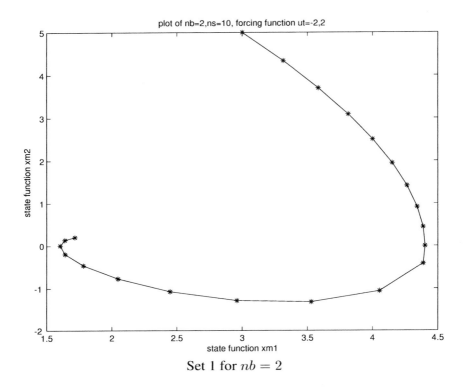

Set 1 for $nb = 2$

The first set of the graphs is the solution at nb (the number of big time subintervals) equals 2. The time period $[0, 1]$ is divided into $[0, t_1]$ and $[t_1, 1]$. ns (the number of small time subintervals) takes $1, 2, 4, 6, 8, 10$ respectively. As mentioned in Section 3.3, the control policy allows the control only to jump at the end points of the big time intervals (so the control does not jump at the end of the small time intervals). The control takes value $-2, 2$ in two big time intervals $[0, t_1], [t_1, 1]$ and switches once at time t_1. The first pair of graphs are the results of $nb = 2$ and $ns = 1$, that is, the time period is divided by 2 big intervals and each subinterval has no further subdivision. It is obvious that the approximation between state and the given fitting function is not very close because of the small nb and ns. As ns (the number of small intervals) increases, a better approximation is obtained. But since control only jumps once during the whole time horizon, it is hard to reach the global minimum. It is understandable that more jumps are helpful for searching a better fit — to find a more stable financial system. Although the ns still increases, the decreasing of the objective function slows down, thus nb needs to be increased. The results are also reported in Table 3.1. These results demonstrate the complexities of the dynamics of the financial system with damped oscillator.

A Financial Oscillator Model

55

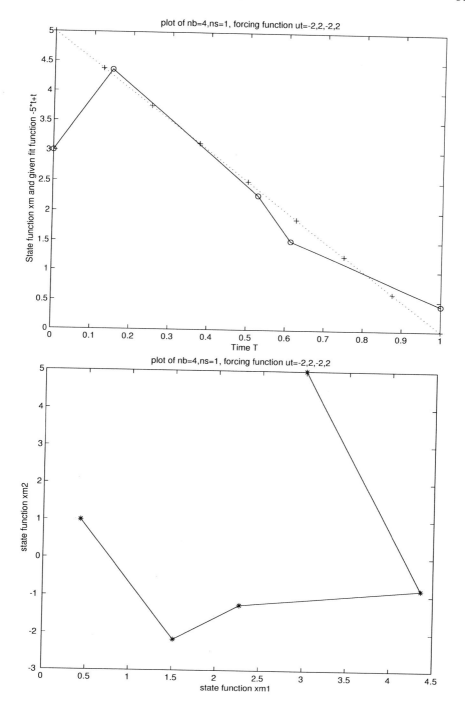

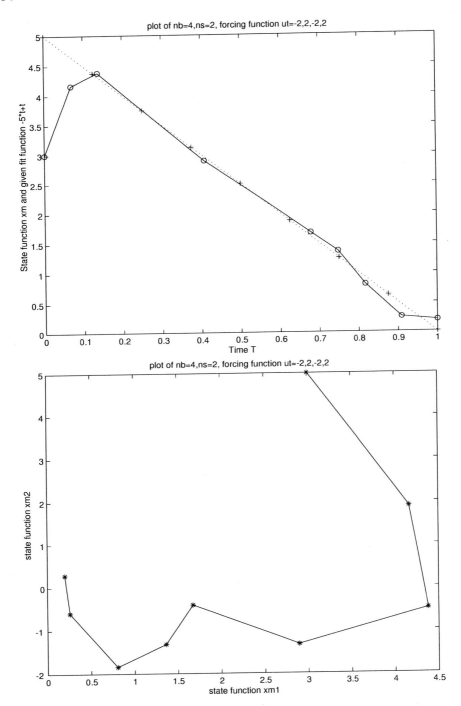

A Financial Oscillator Model

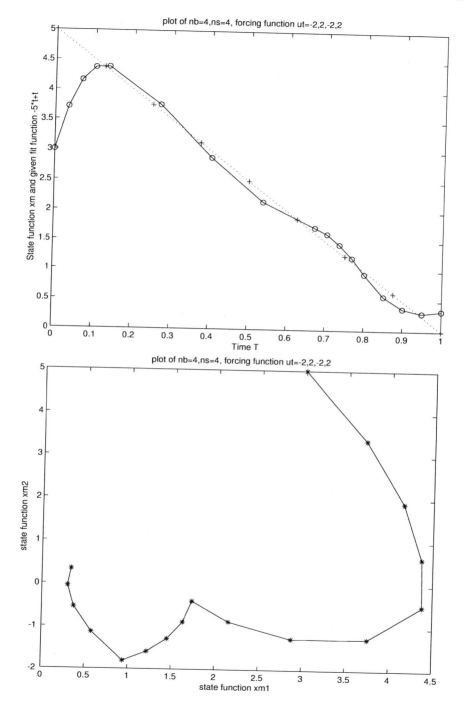

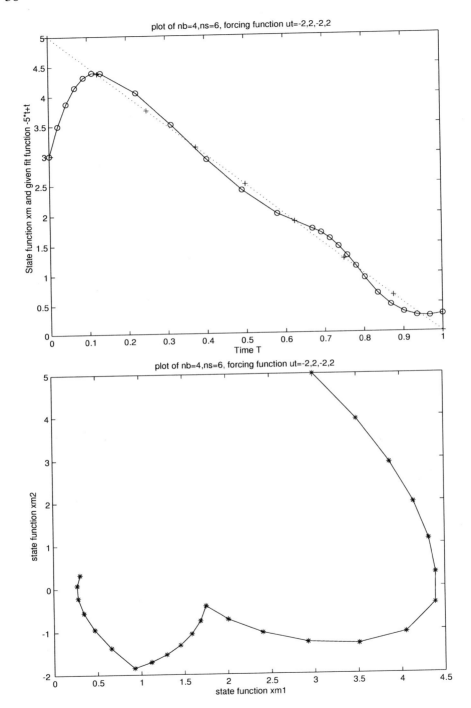

A Financial Oscillator Model

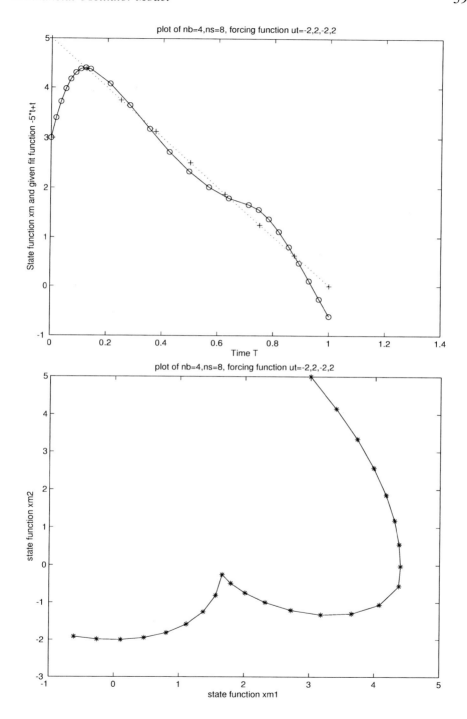

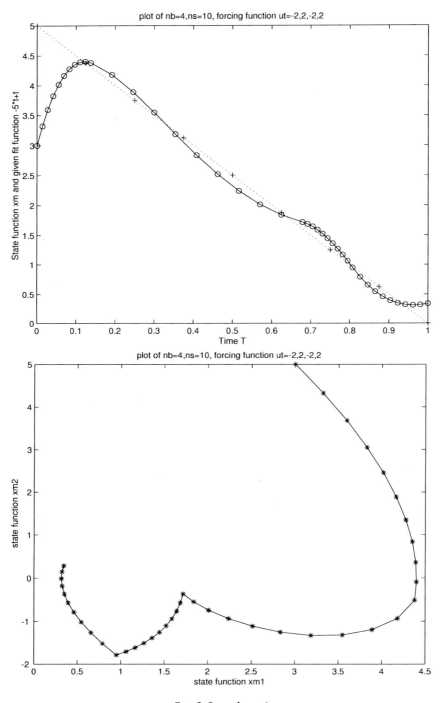

Set 2 for $nb = 4$

A Financial Oscillator Model

This set of the graphs is the graphical solution at $nb = 4$ and $ns = 1, 2, 4, 6, 8, 10$. The control policy is $u(t) = -2, 2, -2, 2$ in successive time intervals $[0, t_1], [t_1, t_2], [t_2, t_3], [t_3, 1]$ and control jumps three times at t_1, t_2, t_3. A better approximation than the results at $nb = 2$ is shown. Most of the results confirm the computational algorithms 3.1-3.4 except a typical case $nb = 4, ns = 8$. In this case, the value of the objective function is bigger than the value of the objective function at $nb = 4, ns = 6$. That means the decrease of the objective function stopped at this point and increased instead. We also discovered that the value of the objective function at this point is same as the value of the other control pattern $2, -2, 2, -2$ at the same point. They are shown in Table 3.1 and Table 3.2. From this test, it may be conjectured that the optimal search stops at a certain point for some reason. A small test was also made in the experiment, another control policy, which was created as $u(t) = -2.05, 2, -2.05, 2$, was put into the program to replace the old control policy. A better search was obtained and the value of the objective function of the financial model became $J = 0.2034$. The new control policy enables the optimal search to continue until the optimum is reached.

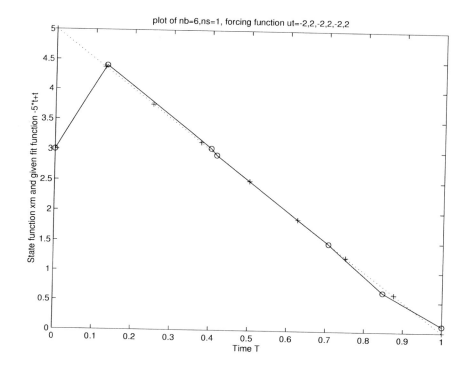

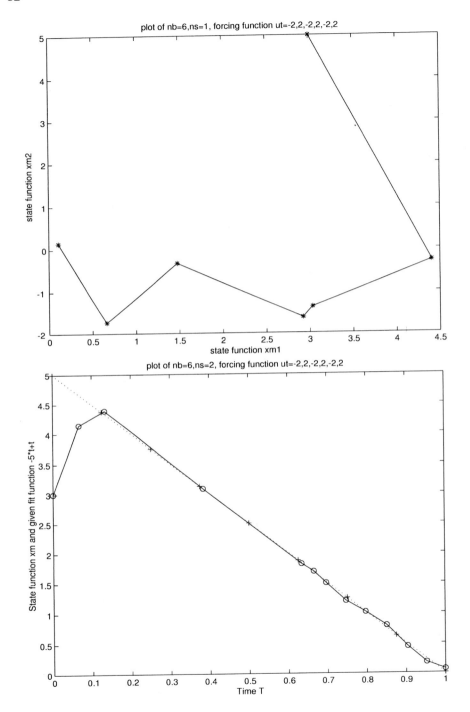

A Financial Oscillator Model

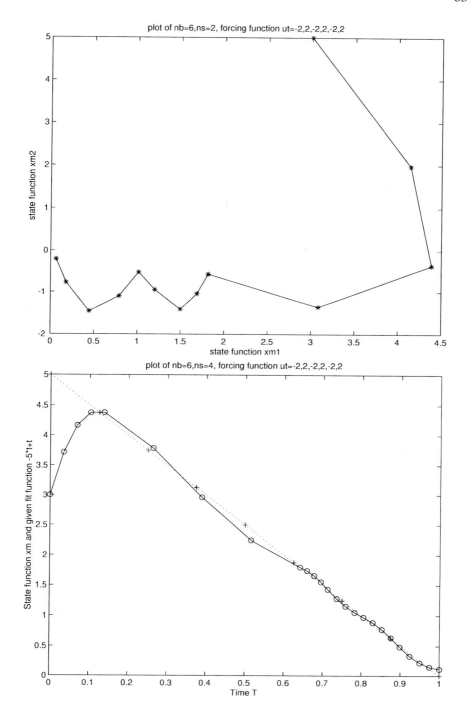

64　　　　　　　　　　　　　　　　*OPTIMAL CONTROL MODELS IN FINANCE*

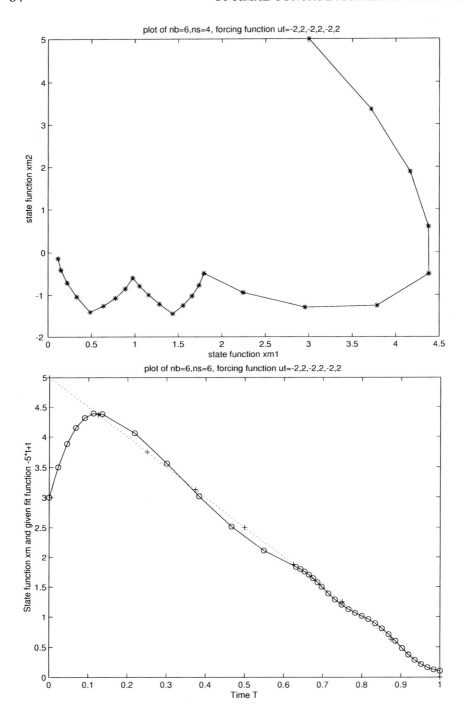

A Financial Oscillator Model

65

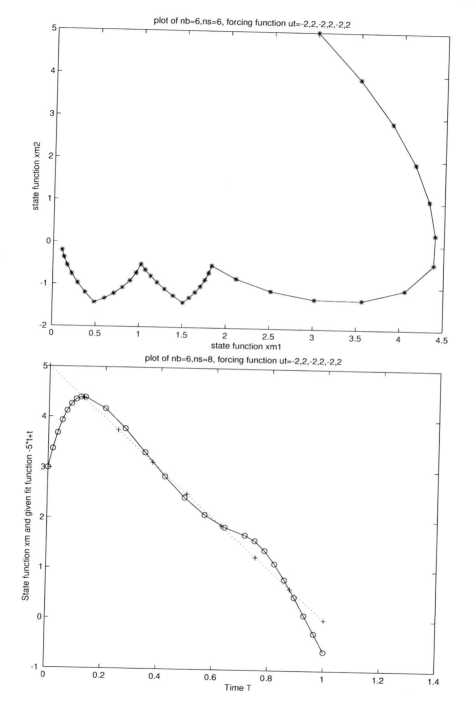

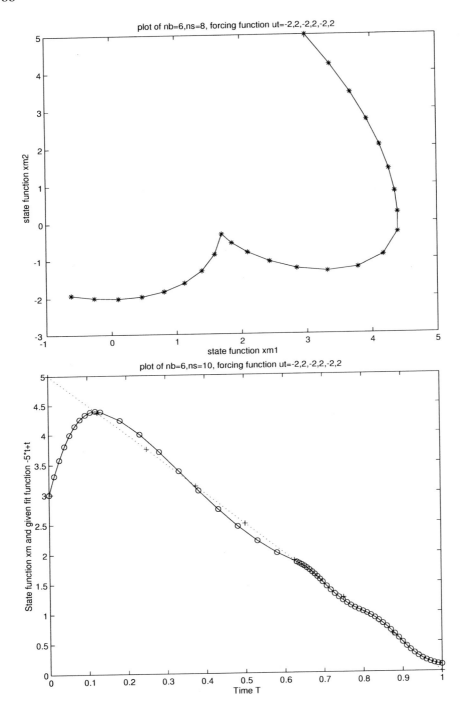

A Financial Oscillator Model

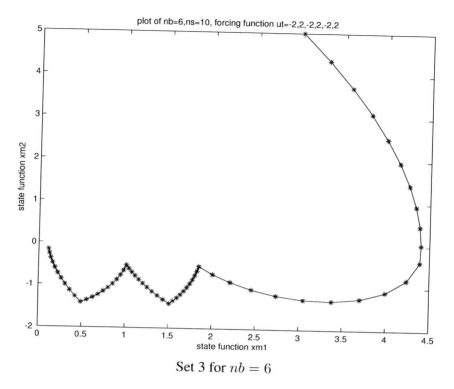

Set 3 for $nb = 6$

This is the set of the graphical results at $nb = 6$ and $ns = 1, 2, 4, 6, 8, 10$. A better approximation than $nb = 4$ is obtained as expected. A similar computation as in case $nb = 4, ns = 8$ happened at case $nb = 6, ns = 8$ and case $nb = 6, ns = 64$. In case $nb = 6, ns = 8$, the optimal switching times are $t_1 = 0.129, t_2 = 0.709, t_3 = 0.999, t_4 = t_5 = t_6 = 0.999$, the control stays at -2 from the third subinterval until the end of the time period. Although the control policy was set to switch five times in the program, the real computation did not show that the control jumped so often. As in the experiment we did for case $nb = 4, ns = 8$, this problem also can be solved by perturbing the control by a small amount. From the results shown in the Table 3.1, a conclusion can be made that when the number of big intervals and small intervals increases, the result of the objective function decreases. Although there are some unexpected results, a small disturbance of the control can easily obtain the correct answers. The experiment gave some promising results which confirm the accuracy of the computational algorithms in this chapter.

Another experiment is also presented to indicate the effects of the different patterns of the control.

68   *OPTIMAL CONTROL MODELS IN FINANCE*

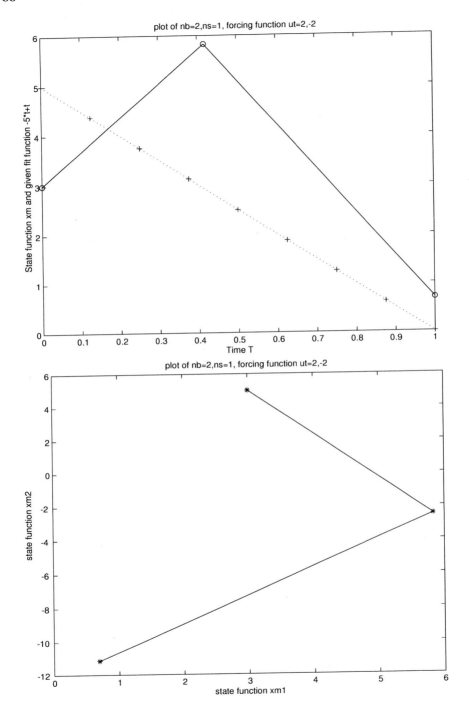

A Financial Oscillator Model

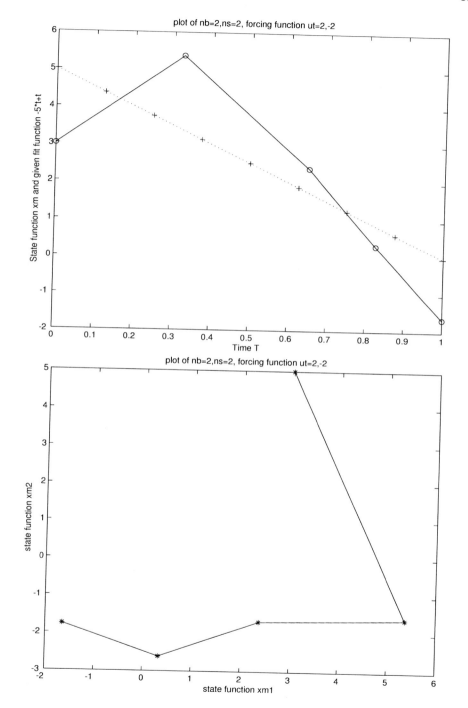

70   OPTIMAL CONTROL MODELS IN FINANCE

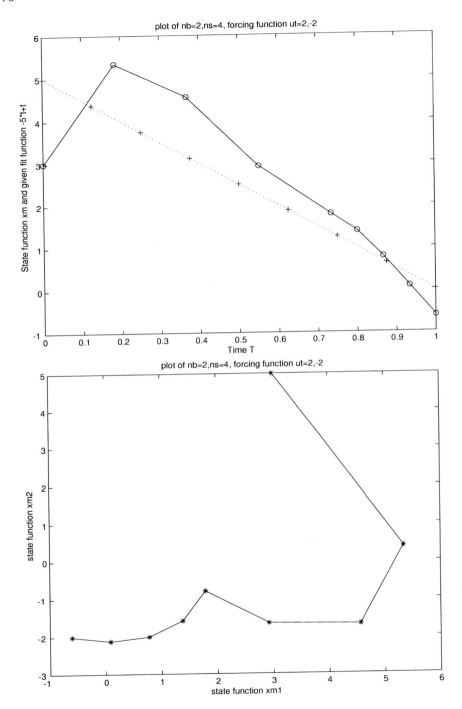

A Financial Oscillator Model

71

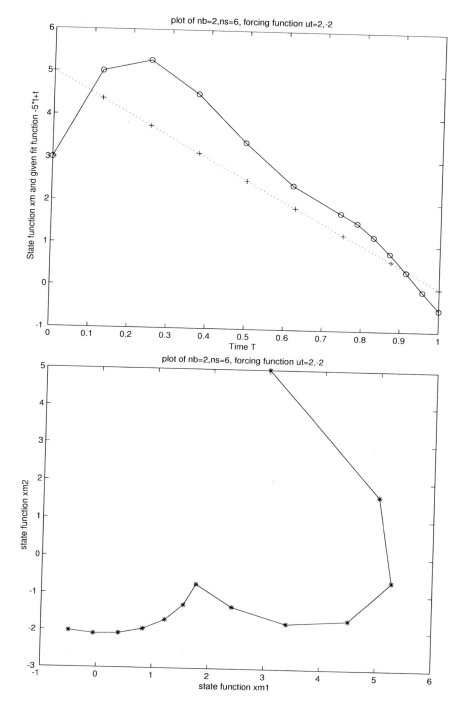

72 OPTIMAL CONTROL MODELS IN FINANCE

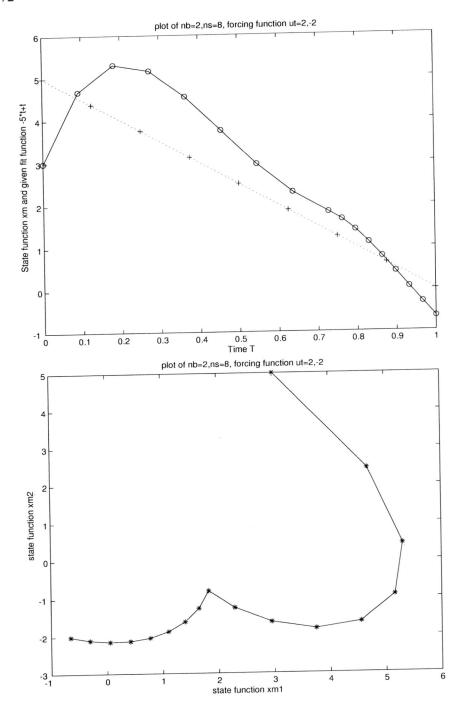

A Financial Oscillator Model

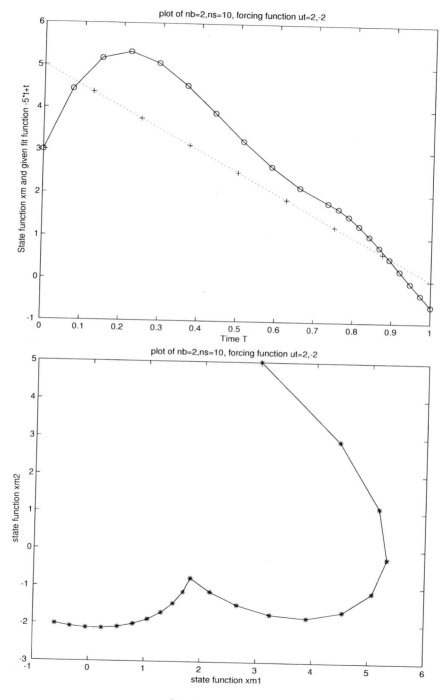

Set 1 for $nb = 2$

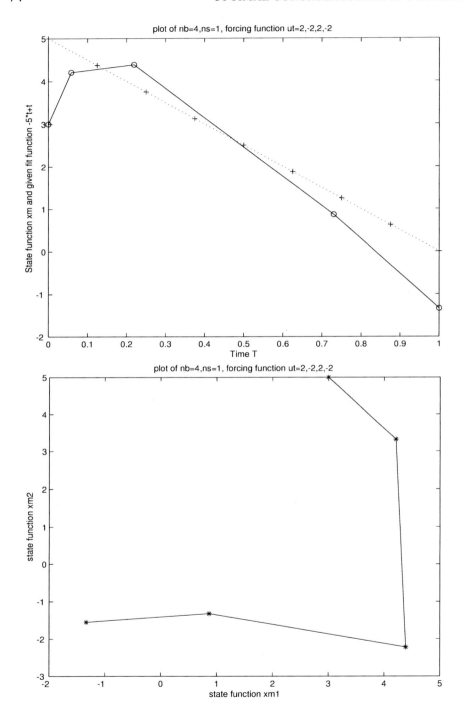

A Financial Oscillator Model

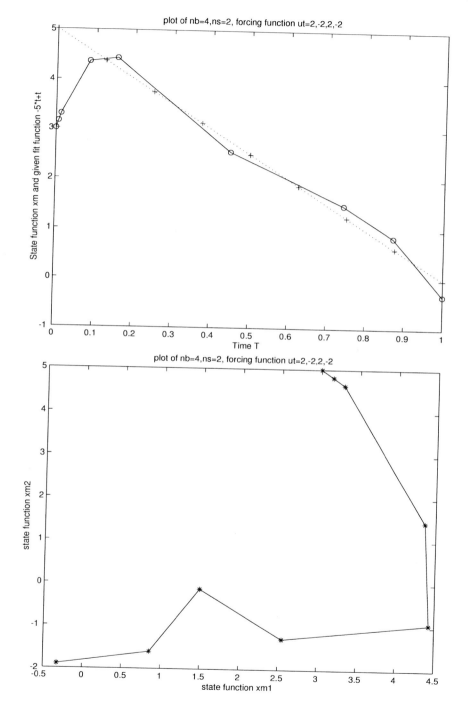

76 OPTIMAL CONTROL MODELS IN FINANCE

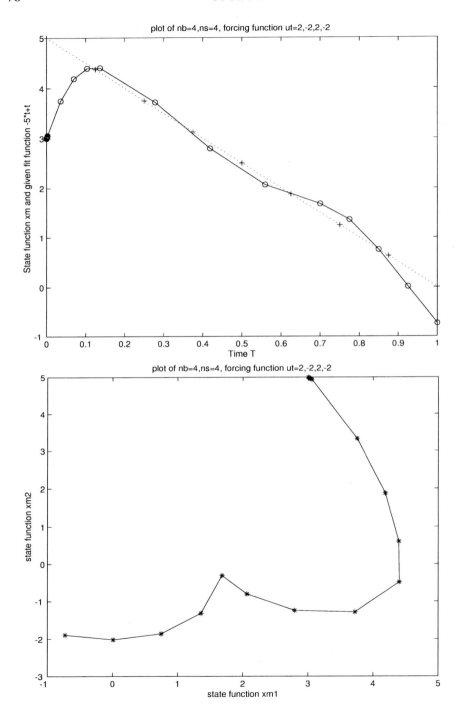

A Financial Oscillator Model

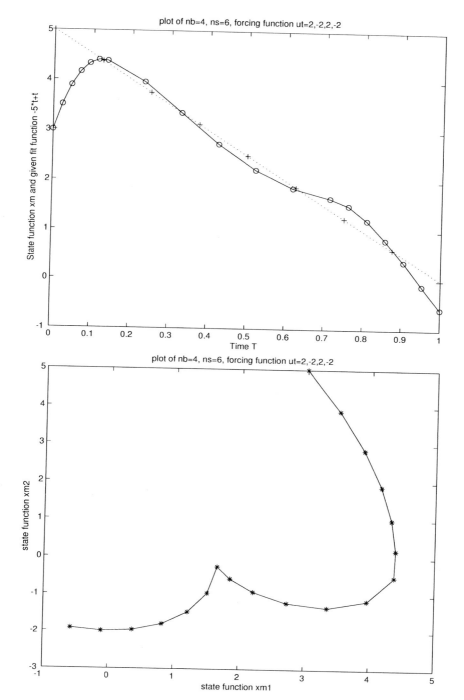

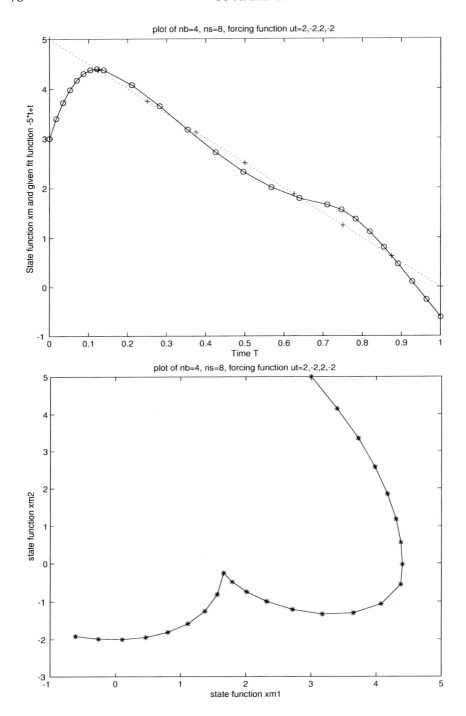

A Financial Oscillator Model

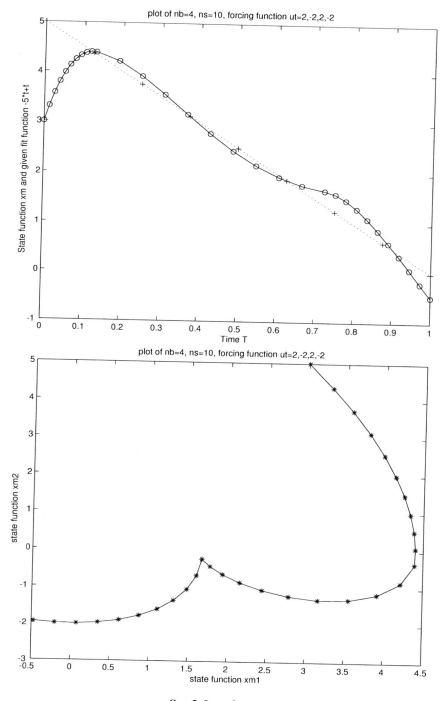

Set 2 for $nb = 4$

80 OPTIMAL CONTROL MODELS IN FINANCE

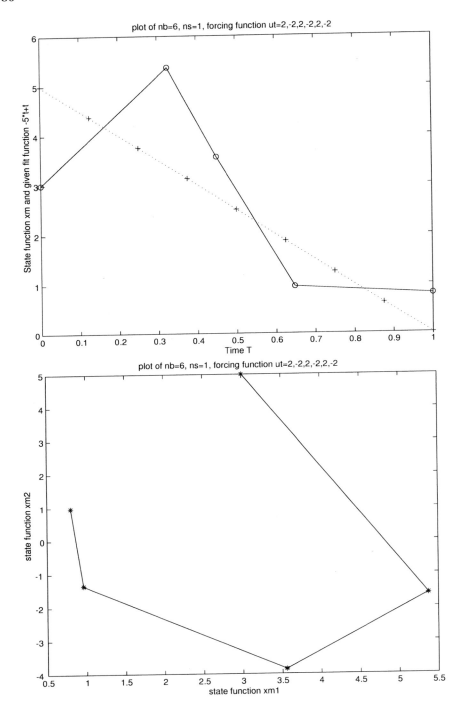

A Financial Oscillator Model

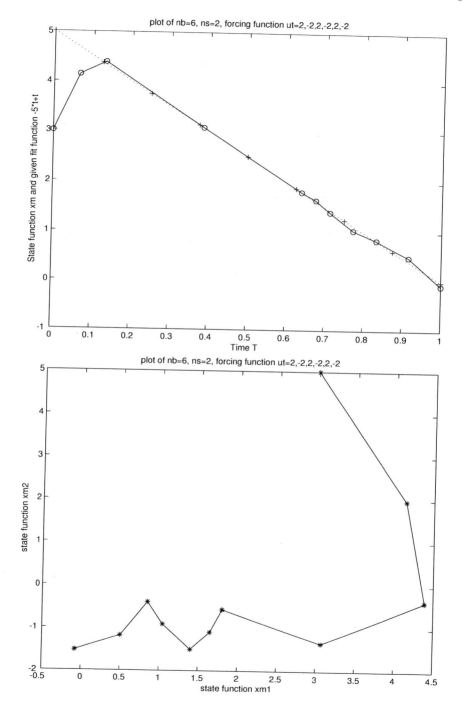

82   OPTIMAL CONTROL MODELS IN FINANCE

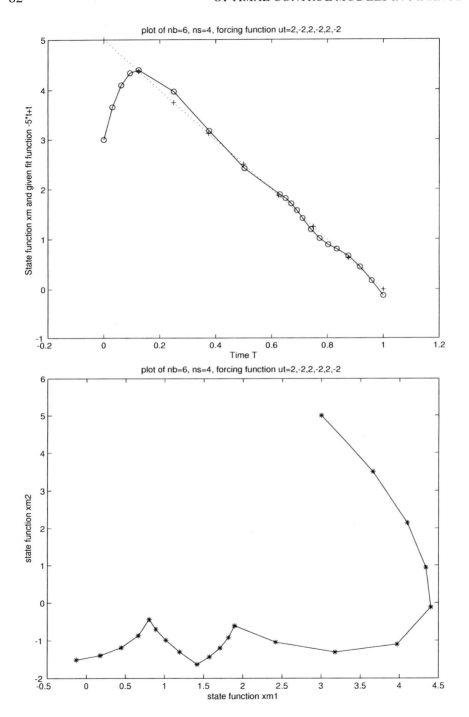

A Financial Oscillator Model

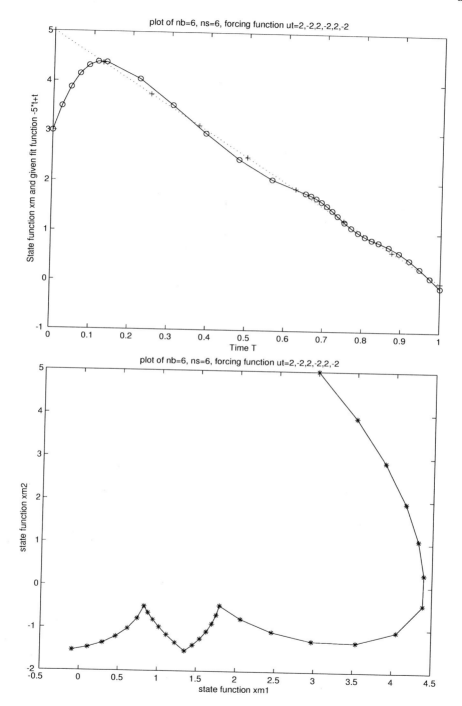

84   OPTIMAL CONTROL MODELS IN FINANCE

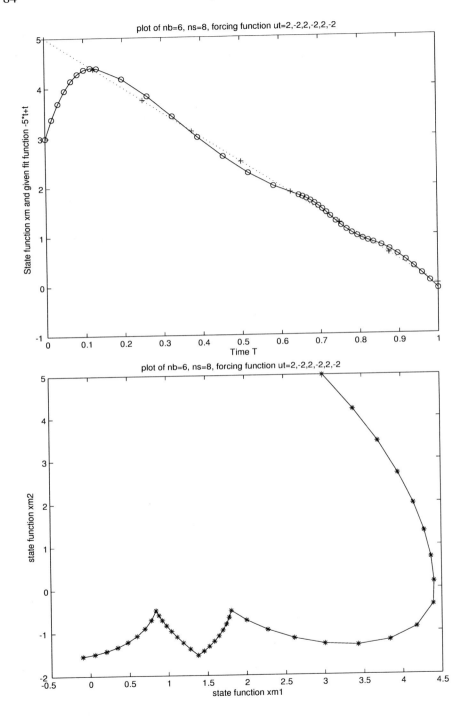

A Financial Oscillator Model

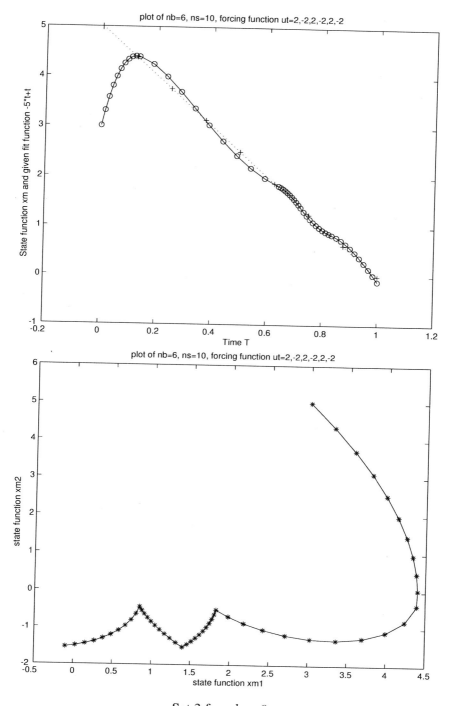

Set 3 for $nb = 6$

Table 3.2. Results of the objective function at control pattern 2,-2, ...

F	$n=2$	$n=4$	$n=6$	$n=8$
ns=1	1.8226	0.6311	0.5289	0.3166
ns=2	0.9394	0.3387	0.2614	0.3361
ns=4	0.8744	0.2786	0.2052	0.1959
ns=6	0.8713	0.2609	0.1904	0.1748
ns=8	0.8434	0.2492	0.1759	0.1684
ns=10	0.8261	0.2460	0.1745	0.1648
ns=16	0.8032	0.2322	0.1624	0.1598
ns=32	0.7913	0.2263	0.1582	0.1488
ns=64	0.7832	0.2246	0.2229	0.1474

In this group of graphs, the pattern of control is $2, -2, 2, -2, \ldots$. Like the previous experiment, this experiment starts from $nb = 2, ns = 1$ and increases nb and ns gradually. As expected, better approximation is obtained by increasing either nb or ns. Unexpected answers were also found at certain points. A small perturbation is helpful to gain the accurate results. Here, we only put the attention on the comparison of these two different patterns of the control. From Table 3.1 and Table 3.2, it is found that the values of the objective function of two different control patterns at starting points ($nb = 2, ns = 1$) have big differences. Pattern $-2, 2, -2, 2, \ldots$ gives much better result than pattern $2, -2, 2, -2, \ldots$. It is an interesting phenomenon that when nb and ns become very big (more jumps and better gradient), the results of the objective function with different control patterns are very close. We can conclude that when optimal control jumps infinitely and the integration calculation has more subdivisions of the time period, the better fit for the problem can be reached whatever pattern of the optimal control is used.

Figure 3.1 and Figure 3.2 show the results of the objective function against $1/ns$ at two different control patterns. The value of the objective function tends to zero when the number nb of subdivisions increases. It is also confirmed that a greater number of subdivisions of the time interval will lead to a better integral calculation. The three lines in each figure also show that more jumps of the control will give better approximation. In the next figure, a cost k is added to the objective function. A complete description of the cost of switching control has been discussed in Chapter 2. In this chapter, a cost is only attached to the number of large subintervals concerned with the control jumping, shown as follows:

$$F = \int_0^1 |x_1(t) + 5t - 5| dt + K * nb \qquad (3.23)$$

K is the cost of changing control, nb is the number of large subintervals. In this computation, K is set to be 0.01, and $ns = 64$ for the accuracy of the

A Financial Oscillator Model

calculation. The same constraints (3.17)-(3.22) are included for this modified optimal control problem (3.23).

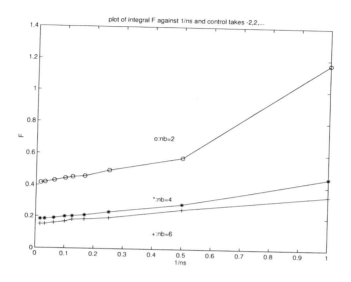

Figure 3.1. Plot of integral F against 1/ns at ut=-2,2

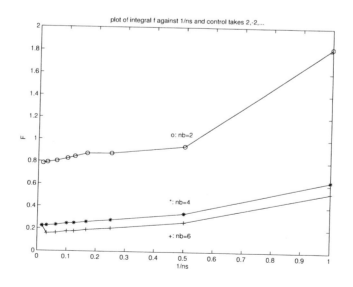

Figure 3.2. Plot of integral F against 1/ns at ut=2,-2

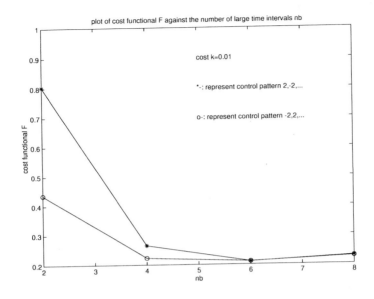

Figure 3.3. Plot of cost function F against the number of large time intervals nb

As discussed earlier, two different levels of the subdivision of the time intervals are used in this example. But the control policy is only constructed so that the control switches at the end of the large subintervals. In two experiments shown in Figure 3.3, a cost $k = 0.01$ attached to nb (the number of the large subintervals) is added to the objective function. It is known that a greater number of subintervals lead the system to a more accurate minimization. A more accurate value of the objective function can be obtained when $ns \to \infty$. The cost is attached to the results of the objective function at $ns = 64$. Two different lines in Figure 3.3 represent the results of the two different patterns of the control which are shown in Figure 3.1 and Figure 3.2. At point $nb = 6$, both of them reach the optimum. It is confirmed that the financial system with two different control patterns will reach the same minimum with the same cost when the number nb of subdivisions increases.

Since there are some non-switching times obtained from the computation, it is conjectured that whether an optimal control with a zero value in the middle to match the non-switching times can bring a better approximation. An example with this matching control will be discussed in Chapter 5.

7. Financial Investment Implications and Conclusion

Since the financial sector is volatile, a financial oscillator model is necessary to study the non-linear complex dynamic behavior of the sector. Incorporation of a damping function to stabilize the oscillatory dynamics of the financial sector can facilitate an understanding of the control mechanism useful for smooth functioning of the financial market. The modeling and computational experiments in this chapter show similar results as those obtained in Chapter 2. Switching times and costs of switching control are significant factors in the determination of optimal investment planning strategy for the economy. Higher costs of switching control reduce the optimal number of switching times essential in evolving an optimal investment strategy overtime. The dampening factor is also significant in designing stable optimal investment planning.

Chapter 4

AN OPTIMAL CORPORATE FINANCING MODEL

1. Introduction

In this chapter, an optimal corporate financing model (Davis and Elzinga [22, 1970]) has been used since it is one of the well known and pioneering models on optimal control in finance. Some other recent work in this area include [21, 1998] and [61, 2001]. The approaches that were constructed in Chapter 2 and Chapter 3 for the optimal control problems are applied. The model discusses investment allocation in order to decide what proportion of its earnings should be retained for internal investment and what proportion should be distributed for shareholders and dividends in a public utility. The aim is to choose the "smart" investment program that the owners can get most benefits from. In the real world, these kinds of problems are common.

Section 4.2 defines the problem of this financial model. Then the analytical solution, which was created by Davis and Elzinga, is discussed in Section 4.3. In Section 4.4, an important technique called "penalty term" is introduced for solving optimal control problems with constraints and end-term condition. By using the "penalty term", all the constraints become easy to be included into the cost function. The transformations of this model are described next in Section 4.5 for the computation. The computational algorithms for this model are constructed in Section 4.6. A computer software package for the algorithms is shown in Appendix A.3. The analysis and discussion of the computing results are presented in Section 4.7.

2. Problem Description

A firm decides how it should generate its finance to maximize the value of the firm, the stock value, or to achieve any other specified objectives. Two well known papers by Modigliani and Miller [59, 1958] [60, 1963] were instrumental in developing the literature on the modern theory of optimal corporate finance. According to this theory, the optimal financial structure of the firm is determined by the optimal financing level, the cost of capital or the weighted average cost of capital (ρ) is equal to the weighted average costs of alternative sources of financing. For a firm, if funds can be obtained from debt or equity and retaining earnings, the optimal financial structure of the firm is given at the marginal investment, where:

$$\rho = (\varrho/v)c + (e/v)r$$

where ϱ = the value of stock, v = the current value of the firm - the value of all outstanding claims against the firm's assets, c = the cost of equity, e = the value of retained earnings, r = the opportunity cost of the retained earnings.

It is academically interesting and practically useful to determine the optimal proportion between different funds that minimize the cost of capital and to maximize the value of the firm. In a dynamic framework, there may be switches among funds over time depending on the developments in the firm and in the financial market and the economy. It is, therefore, important to know the optimal timing for switching from one source of funds to another, which is optimal for the firm in minimizing the cost of capital and maximizing the value of firm. Unlike Modigdiani and Miller, it is assumed in this paper that the structure of capital has impact on the values of shares and the firm.

Modeling optimal corporate financial structure is a useful area for research in corporate finance since such models can provide information about the optimal proportion of sources of finances of the corporation, its investment and dividend strategies over a period of time. Dynamic optimization models in the form of optimal control in addressing optimal corporate financial structure have been developed initially by Davis and Elzinga [22, 1970], Krouse and Lee [49, 1973], Elton and Gruber [26, 1975] and Sethi [78, 1978] (see also Craven [14, 1995], Sethi and Thompson [79, 2000]).

A general limitation of the existing literature on optimal corporate financing is that although there are some analytical studies in this area, computational exercises (numerical model building, application of an algorithm and development or application of a computer program) are not well known (except Islam and Craven [38, 2002]). One limitation of the existing literature on the computation of optimal corporate financing models (including Islam and Craven [38, 2002]) is that the algorithms applied to solve these models produce optimal levels of various funds, but do not generate optimal timing for switching from one fund to another fund. In this paper, a computational approach (algorithm and program) which can generate optimal switching time among different funds is presented.

An Optimal Corporate Financing Model

In terms of algorithm, there is a scope for improvement in the existing algorithms for computing optimal corporate financial models as well. The limitation of the existing literature on optimal control with switching times is that computation of such models is performed by algorithms based on discretization of the switching time. After switching time discretion, the resulting model is a discrete time continuous variable optimal control model with switching times. If the switching times are sub-division times of time discretization, this method can generate a unique solution. In models where switching time is made a parameter or a variable, the determination of optimal switching time may also be difficult due to the difficulty of the switching times and time sub-divisions not coinciding. If the sub-division times and the switching times are different, then the computation involves two steps. First, to find the optimal solution for the time sub-division and later to find optimal solution in terms of optimal switching time. Computing an optimal financial model with too many time steps may be difficult in some cases due to computer memory and time required to compute.

A computational approach, which can overcome the above computational problems by suitable transformation of the original time is developed in this chapter.

A simplified non-linear optimal control model which can address the above optimal financing problems for corporations is described below, involving two state variables and two control variables. Price $P(t)$ and equity per share $E(t)$ are the state variables, and the earnings retention rate u_r and stock financing rate u_s are the control variables. The objective of the utility is defined as the discounted sum of dividends and capital gains, thus, the present value of share ownership is to be maximized. Two differential equations describe the change in stock price and equity per share.

The state, control variables, and parameters are expressed as follows:

State variables
$P(t)$ = market price of a share of stock
$E(t)$ = equity per share of outstanding common stock (net worth of utility divided by the number of shares outstanding)

Control variables
u_r = retention rate which describes the fraction of earnings retained for increasing the capital assets
u_s = stock financing rate concerning new money invested in the company

Parameters
ρ = market capitalization rate (or investor discount rate)
k = maximum investment rate
r = rate of return to equity (maximum return allowed by government)

δ = discount on share price resulting from flotation cost
c = a positive constant denoting the responsiveness of the price to changes in earnings and dividends
T = planning horizon of the optimal financing program
$\dot{x}(t) = dx(t)/dt$

The objective function of this model is to maximize the present value of the owners' shares. Expressed by an integral plus an end-point term:

$$\text{MAX} P(T)e^{\rho T} + \int_0^T e^{-\rho t}[1 - u_r(t)]rE(t)dt \tag{4.1}$$

subject to:

$$\dot{P}(t) = c[\{1 - u_r(t)\}rE(t) - \rho P(t)] \tag{4.2}$$

$$\dot{E}(t) = rE(t)[u_r(t) + u_s(t)\{1 - \frac{E(t)}{(1-\delta)P(t)}\}] \tag{4.3}$$

$$u(t) \in U = u \mid u_r + u_s \leq k/r < 1, u_r \geq 0, u_s \geq 0 \tag{4.4}$$

where, $P(0) = P_0, E(0) = E_0$ and the terminal condition is given by the fixed planning horizon T.

Davis and Elzinga used the reverse-time construction technique which was given in reference [35, 1965], which will be described below, to systematically construct a complete solution from solution cases. The solution cases are classified according to the solution of a mathematical programming problem of each instant at time arising from the Maximum Principle [69, 1962].

3. Analytical Solution

Davis and Elzinga used the reverse-time construction technique that is particularly significant in solution synthesis. Starting from the terminal manifold and moving backwards in time, the entire space of state is filled with optimal trajectories solving the problem for arbitrary initial states. A linear program solved by inspection originally determined the solution of this model. The Maximum principle was successfully used and also modified to allow ease in handling discounted objective functional. The existence of the optimal control for this problem was established by corollary 2 of Theorem 4 in reference [50, 1967]:

Theorem 4. Consider the non-linear process in R^n:

$$(S)\ \dot{x} = f(x, t, u) \text{ in } C^1 \text{ in } R^{n+1+m}$$

The data are as follows:

1. The initial and target sets $X_0(t)$ and $X_1(t)$ are non-empty compact sets varying continuously in R^n for all t in the basic prescribed compact interval $\tau_0 \leq t \leq \tau_1$.

2. The control restraint set $\Omega(x,t)$ is a non-empty compact set varying continuously n R^m for $(x,t) \in R^n \times [\tau_0, \tau_1]$.

3. The state constraints are (possibly vacuous) $h^1(x) \geq 0, \ldots, h^r(x) \geq 0$, a finite or infinite family of constraints, where $h^1, \ldots, h^r$ are real continuous functions on R^n.

4. The family $\mathcal{F}$ of admissible controllers consists of all measurable functions $u(t)$ on various time intervals $t_0 \leq t \leq t_1$ in $[\tau_0, \tau_1]$ such that each $u(t)$ has a response $x(t)$ on $t_0 \leq t \leq t_1$ steering $x(t_0) \in X_0(t_0)$ to $x(t_1) \in X_1(t_1)$ and $u(t) \in \Omega(x(t), t), h^1(x(t)) \geq 0, \ldots, h^r(x(t)) \geq 0$.

5. The cost for each $u \in \mathcal{F}$ is:

$$C(u) = g(x(t_1)) + \int_{t_0}^{t_1} f^0(x(t), t, u(t)) dt + \max_{t_0 \leq t \leq t_1} \gamma(x(t))$$

where $f^0 \in C^1$ in R^{n+1+m}, and $g(x)$ and $\gamma(x)$ are continuous in R^n.

Assume:

a. The family $\mathcal{F}$ of admissible controllers is not empty.

b. There exists a uniform bound:

$$|x(t)| \leq b \text{ on } t_0 \leq t \leq t_1$$

for all responses $x(t)$ to controllers $u \in \mathcal{F}$.

c. The extend velocity set:

$$\dot{V}(x,t) = f^0(x,t,u), f(x,t,u) | u \in \Omega(x,t)$$

is convex in R^{n+1} for each fixed (x,t).

Then there exists an optimal controller $u^*(t)$ on $t_0^* \leq t \leq t_1^*$ in $\mathcal{F}$ minimizing $C(u)$.

Corollary 2. Consider the control process in R^n:

$$(S) \quad \dot{x} = A(x,t) + B(x,t)u$$

with cost:

$$C(u) = g(x(t_1)) + \int_{t_0}^{t_1} [A^0(x(t), t) + B^0(x(t), t)u(t)] dt + \text{ess.sup } \gamma(x(t), u(t))$$

where:
$$t_0 \leq t \leq t_1$$

where the matrices A, B, A^0, B^0 are C^1 functions on R^{n+1}, $g(x)$ and $\gamma(x, u)$ are continuous in R^{n+m}, and $\gamma(x, u)$ is a convex function of u for each fixed x. Assume that the restraint set $\Omega(x, t)$ is compact and convex for all (x, t). Then hypothesis c. is valid. If we assume 1. to 4. and a. b., then the existence of an optimal control $u_0^*(t)$ on $t_0 \leq t \leq t_1^*$ in $\mathcal{F}$ is assured.

Due to the Maximum Principle, the necessary conditions for $u(t)$ to be optimal are:

$$\text{MAX}_{u \in U} H = \psi_0 rEe^{-\rho t} + c\psi_P[rE - \rho P] + rE[\psi_E - c\psi_P - \psi_0 e^{-\rho t}]u_r$$
$$+ rE\psi_E[1 - \tfrac{E}{(1-\delta)P}]u_s$$

where:

$$\psi_0 = \text{constant} \geq 0$$

$$\dot{\psi}_P = c\rho\psi_P - (E/P)u_s \frac{\psi_E rE}{(1-\delta)P}$$

$$\dot{\psi}_E = r(c\psi_P + \psi_0 e^{\rho t}) - r[\psi_E - c\psi_P - \psi_0 e^{\rho t}]u_r$$
$$- r\psi_E[1 - \tfrac{E}{(1-\delta)P}]u_s + [\tfrac{\psi_E rE}{(1-delta)P}]u_s$$

$$\psi_P(T0) = \psi_0 e^{-\rho T}$$

$$\psi_E(T) = 0$$

$$\psi = (\psi_0, \psi_P, \psi_E) \neq 0$$

Here, $\psi_0 = 1$. The Hamiltonian can be modified by introducing new "steady state" variables $\lambda_P = \psi_P e^{\rho t}, \lambda_E = \psi_E e^{\rho t}$.

$$\dot{H} = e^{-\rho t} H = \dot{H}(\lambda, P, E, u) = h(\lambda, P, E) + S_r(\lambda, P, E)u_r + S_s(\lambda, PE)u_s \tag{4.5}$$

$$h(\lambda, P, E) = rE + r\lambda_P(rE - \rho P)$$

$$S_r(\lambda, P, E) = rE[\lambda_E - c\lambda_P - 1]$$

$$S_s(\lambda, P, E) = rE\lambda_E[1 - \frac{E}{(1-\delta)P}]$$

An Optimal Corporate Financing Model

The adjoint variables are defined as follows:

$$\dot{\lambda}_P = (c+1)\rho\lambda_P - (E/P)\frac{\lambda_E Er}{(1-\delta)P}u_s \qquad (4.6)$$

$$\begin{aligned}\dot{\lambda}_E &= -r(c\lambda_P + 1) + \rho\lambda_E - r(\lambda_E - c\lambda_P - 11)u_r \\ &\quad -r\lambda_E[1 - \frac{E}{(1-\delta)P}]u_s + \frac{\lambda_e Eru_s}{(1-delta)P}\end{aligned} \qquad (4.7)$$

$$\dot{P} = c[Er(1 - u_r) - \rho P] \qquad (4.8)$$

$$\dot{E} = Er[u_r + u_s(1 - \frac{E}{(1-\delta)P})] \qquad (4.9)$$

$$U = u|u_r + us \leq k/r, u_r \geq 0, u_s \geq 0 \qquad (4.10)$$

with boundary conditions $P(0) = P_0, E(0) = E_0, \lambda_P(T) = 1, \lambda_E(T) = 0$. Parameterizing terminal values of the state variables as $P(T) = s_P > 0$ and $E(T) = s_E > 0$.

The maximization of H with respect to u can be characterized as follows:

(a) $S_r > 0, S_r > S_s, u_r^* = k/r, u_s^* = 0$
(b) $S_r > 0, S_s > S_r, u_r^* = 0, u_s^* = k/r$
(c) $S_r < 0, S_s < 0, u_r^* = 0, u_s^* = 0$
(d) $S_r = 0, S_s < 0, 0 \leq u_r^* \leq k/r, u_s^* = 0$
(e) $S_r < 0, S_s = 0, u_r^* = 0, 0 \leq u_s^* \leq k/r$
(f) $S_r = S_s > 0, u_r^* \geq 0, u_s^* \geq 0, u_r^* + u_s^* = k/r$
(g) $S_r = 0, S_s = 0, u_r^* \geq 0, u_s^* \geq 0, u_r^* + u_s^* = k/r$

The synthesis of the solution was done by a construction technique in the reverse time sense. The complete solution is given as follows:

In the case (a) $\quad \frac{1}{1-\delta} < r/\rho < \frac{c+[c+(1-\delta)+1](k/\rho)}{c(1-\delta)}$
A. $u_r^* = 0, u_s^* = 0$.   1. $S_r = 0, S_S < 0$.
B. $u_r^* = k/r, u_s^* = 0$.   2. $S_r = S_s > 0$.
C. $u_r^* = 0, u_s^* = k/r$.   3. $S_r < 0, S_s = 0$.

The solution shows the classical bang-bang control. Although the singular arc cases appear in the synthesis, none of them are optimal. The computational methods established in Chapter 2 and Chapter 3 will be used to verify this solution in Section 4.6.

4. Penalty Terms

A substantial class of optimal control problems will deal with the terminal constraint as well as other constraints that may describe physical limitations on some process. For computational reasons, some penalty terms are required to be used to replace these constraints in order to obtain an unconstrained problem. There are some approaches with good reputations. Since the financial model discussed in this chapter has an end-point constraint on the market price of a share of stock at the end of the time planning horizon, the penalty term is used in the terminal constraint transformation. First an approach is introduced to deal with a minimization problem subject to both inequality and equality constraints:

$$\text{MIN}_x f(x) \text{ subject to } g(x) \leq a, h(x) = b, \text{ where } x \in R^n \quad [A]$$

Consider the objective function $f(.)$ as a cost to be minimized; then additional penalty costs are added to $f(.)$ when x does not satisfy the constraints. Define vector $v_+ = (v_{1+}, v_{2+}, \ldots, v_{n+}) = v = (v_1, v_2, \ldots, v_n)$ if $v_1, v_2, \ldots, v_n \geq 0$, 0 if $v_1, v_2, \ldots, v_n < 0$. The problem (1) is replaced by the following unconstrained problem:

$$\text{MIN}_x f(x) + \frac{1}{2}\mu \parallel [g(x) - a + \mu^{-1}\xi]_+ \parallel^2 + \frac{1}{2}\mu \parallel [h(x) - b + \mu^{-1}\sigma] \parallel^2$$

The terms in g and h consist of the penalty functions. They are zero when x satisfy all the constraints. μ is a positive parameter, that can more generally be replaced by different parameters for each component of $[g(x)]_+$ and $h(x)$. ξ and σ are Lagrange multipliers. From the theory of augmented Lagrangian [14, 1995], this unconstrained problem with the penalty terms is minimized at the same point as the given constrained problem [A], provided that the Lagrange multipliers ξ and σ are suitably chosen.

Before the penalty method of terminal constraints is stated, the Delta function should be introduced since it will be used to include the terminal constraints into the integral later. The Dirac delta function $\delta(.)$ is described by:

$$\delta(.) \geq 0, \delta(t) = 0 \text{ when } t \neq 0 \text{ in } R, \text{ and } \int_R \delta(t)dt = 1.$$

Now consider an optimal control problem of the form:

$$\text{MIN}_{x(.),u(.)} F^0(x,u) := \int_0^T f(x(t), u(t), t)dt + \Phi(x(T))$$

A terminal constraint $\sigma(x(T)) = b$ can be replaced by a penalty term added to $F^0(x,u)$:

$$F(x,u) := F^0(x,u) + \frac{1}{2}\mu \parallel \Phi(x(T)) - b^* \parallel^2$$

where μ is a positive parameter, and b^* approximates to b, thus $b^* = b + \theta/\mu$, and where θ is the Lagrange multiplier.

Thus an "end-point term" in a control problem can be included in the integrand as shown:

$$\int_0^T f(x(t), u(t), t)dt + \Phi(x(T)) = \int_0^T [f(x(t), u(t), t) + \Phi(x(t))\delta(t-T)]dt$$

and so is not an additional case to be treated separately.

A constraint such as $\int_0^1 \theta(u(t)) \leq 0$, which involves controls at different times, can be treated by adjoining an additional state component:

$$y_0(0) = 0, \dot{y}_0(t) = \theta(u(t))$$

and imposing the state constraint $y_0(1) \leq 1$. The latter can be handled by a penalty term $\frac{1}{2}\mu \parallel [y_0(1) - c]_+ \parallel^2$, where the parameter $c = -\theta/\mu$ is small.

5. Transformations for the Computer Software Package for the Finance Model

In this section, in order to develop a computer software package of the optimal control for this financial model, some transformations of the formulas are introduced here to meet the requirement of the computation. Basically most transformation techniques with the division of time scales used here were introduced in Chapter 3. Only the different transformations for this particular computer package are introduced here.

The differential equations in (4.2) and (4.3) are represented as follows:

$$\dot{P}(\tau) = c[\{1 - u_r(\tau)\}rE(\tau) - \rho P(tau)] * nb * (pt_{j+1} - pt_j) \qquad (4.11)$$

$$\dot{E}(\tau) = rE(t)[u_r(t) + u_s(t)\{1 - \frac{E(t)}{\{(1-\delta)P(t)\}}] * nb * (pt_{j+1} - pt_j) \quad (4.12)$$

The integral in (4.1) is transformed as:

$$J = \sum_{i=0}^{r-1} J_i = \sum_{i=0}^{r-1} \int_{ih}^{(i+1)h} nn * e^{-\rho t}[1 - u_r(t)]rE(t) * (t_{j+1} - tj)dt, \quad (4.13)$$

where $h = 1/nn$

Terminal state is treated separately for nqq package:

$$END = P(nn+1) * e^{\rho T} \quad (4.14)$$

Since in this model, time horizon parameter T is assumed to be greater than 1, another transformation of changing time interval $[0, T]$ to $[0, 1]$ is required for the program which only deals with the time period $[0, 1]$.

This transformation was indicated in section 2.3. Let a new time t equal $T * \tau$, here τ is the scaled time set for the computational methods.

Then the differential equations and integral of objective function become:

$$\dot{P}(\tau) = T * (c[\{1 - u_r(\tau)\}rE(\tau) - \rho P(tau)] * nb * (pt_{j+1} - pt_j)) \quad (4.15)$$

$$\dot{E}(\tau) = T * (rE(t)[u_r(t) + u_s(t)\{1 - \frac{E(t)}{\{(1-\delta)P(t)\}}] * nb * (pt_{j+1} - pt_j)) \quad (4.16)$$

The integral in (4.1) is transformed as:

$$J = \sum_{i=0}^{r-1} J_i = \sum_{i=1}^{r-1} \int_{ih}^{(i+1)h} T * (nn * e^{-\rho tT}[1 - u_r(t)]rE(t) * (t_{j+1} - tj)dt, \quad (4.17)$$

where $h = 1/nn$

An Optimal Corporate Financing Model

The transformations here only deal with the time variable t. The end-point term is considered separately in the program.

6. Computational Algorithms for the Non-linear Optimal Control Problem

From section 4.3, we know that at a particular case (a), the bang-bang control is the optimal solution of the system. $u(t) := (u_r(t), u_s(t))$ lies on the vertices of a triangular area:

$$u_r(t) \geq 0, u_s(t) \geq 0; \tag{4.18}$$

$$u_r(t) + u_s(t) \leq k/r < 1. \tag{4.19}$$

However, a *singular arc* is possible, with $u(t)$ lying on an edge of the triangle (instead of a vertex), when the parameters (p, c, r, etc.) of the functions take particular values. Since Davis and Elzinga did not have a computer software package for this optimal control problem, this research is focused on developing computer software based on the research work which has been done in Chapter 2 and Chapter 3. In order to get accurate estimates of the switching times, some computational methods have to divide $[0, T]$ into many subintervals. The time transformation introduced by Goh and Teo [33, 1987] makes it possible to avoid this difficulty. The technique was also used to construct the computational algorithms in this section. We establish the control policy as bang-bang control here. The optimal switching times will be computed. Several sequences of the control patterns will be experimented with different initialization of the states. The results will be discussed and analyzed in next section.

Computational method 4.1 Main Model Program (see model1_1.m in Appendix A.3)

Step1. Initialization. First set the a vector of $parameters = [p, k, r, d, c, t]$ which includes all the parameters in this financial model. $p = $ is market capitalization rate, $k = $ maximum investment rate, $r = $ rate of return to equity, $d = $ discount on share price resulting from flotation costs, $c = $ a positive constant denoting the responsiveness of the price to changes in earnings and dividends, $T = $ planning horizon of the capital budgeting program. Then set parameter $par = $ [the number of the state components, number of control components, $nb = $ the number of big subintervals, $ns = $ the number of small subintervals, $parameters$]; and get the total number of the subintervals by calculating $nn = nb * ns$. Set the MATLAB "constr" function parameters $par(13) = 1$ (one equation constraint in the minimization problem), $par(14)$ = the maximum number of function evaluations, and arbitrary starting lengths of the switching time intervals $um0 = (um0_1, um0_2, ..., um0_{nb})$. Set the vectors of upper bounds uu and the vector of lower bounds ul of um, thus $uu \leq um \leq ul$. Also set the initial state $xinit$;

Step2. Call the MATLAB "constr" function. In turn, "constr" calls the "Model2" to calculate the minimization of the calling program with respect to the optimal vector um;

Step3. Input the optimal result um to "Model2", to obtain the values of the objective function $J(nn)$ (the last value of the integral) and state vector xm (xm is a vector of all the values of the state functions take at the gridpoints of the switching time intervals). The result of the objective with respect to optimal switching times is the solution of this financial system model.

Computational method 4.2: Model 2 (see model1_2.m in Appendix A.3)

Step1. Initialization. Input um, par and initial state $xinit$. Set the initial state $xm(1,:) = xinit$, and $nx = par(1)$, the number of the state components, $nu = par(2)$, the number of the control components (in this case, nx and nu both equal 2, there are two states and two controls,) $nb = par(3)$, the number of big time intervals, $ns = par(4)$, the number of small time intervals, $nn = par(3)*par(4)$, the number of total time intervals, initial scaled time $t = 0$, subinterval counter $it = 1$, hs = length of the whole subintervals. Choose the "Model3" as the right side of the differential equations (4.2) and (4.3), input um;

Step2. Construct the vector sm whose components represent the lengths of the total time intervals by dividing each big time interval um by ns;

Step3. Call the SCOM package function "nqq" with the stated "Model3" to solve dynamic equations (4.2) and (4.3). Tabulate the solution for the state as the vector $xm(:1) = [x_1(1), x_1(2), ..., x_1(nn)]$, the result of (4.2), $xm(:2) = [x_2(1), x_2(2), ..., x_2(nn)]$, the result of (4.3);

Step4. Set the initial scaled time $t = 0$, subinterval counter $it = 1$, initial state $zz(:1) = 0$, $ma = 1$, the number of the input state function. Choose the "Model3" for SCOM function "nqq". Input the vector $xm(:1)$ and um;

Step5. Call SCOM function "nqq" with the stated "Model4" to solve the differential equation $d(J(w(t)))/dt = ...$ Tabulate the results in $w(.)$ as the components of the vector $jm = [j(1), ..., j(nn)]$;

Step6. Obtain the last value $xm(nn, 1)$ of the state vector for the "end-point condition";

Step7. Call the "End-point condition" to calculate the terminal state;

Step8. Add the result gained from "End-point" condition to the last value of the vector jm as the result of the objective function $w(t)$ in (4.1), and calculate the constraint function of 'model2" which is $g(um) = \sum_{i=1}^{N} um_i - 1$.

Computational method 4.3: Model 3 (see model1_3.m in Appendix A.3)

Step1. Initialization. Input scaled time t, subinterval counter it, the length of total subintervals hs, vector um and vector par. Set the value of parameters p, k, r, d, c, T. Set the number of the total subintervals $nn = 1/hs$;

Step2. Obtain the number of the big time intervals by $nb = nn/ns$ for constructing the optimal control policy;

Step3. Set the control policy as vector $u = [u_1, u_2]$ which jumps between the vertices of the triangle area (4.18)-(4.19). The control only jumps at the end points of the big time intervals;

Step4. Construct the right side of the transformation equation for time pt in (3.10);

Step5. Obtain the right side of the differential equations by using the transformations in (4.15), (4.16).

Computational method 4.4: Model 4 (see model1_4.m in Appendix A.3)

Step1. Initialization. Input scaled time t, subinterval counter it, the length of subintervals hs, vector xm of values of first state function at switching times, and also the initial state z and vector sm. Set total subintervals $nn = 1/hs$;

Step2. Use linear interpolation to get the estimate "xmt" of the state, in a time t between grid points $0h, h, 2h, \ldots, h*nn$, where $h = 1/nn$;

Step3. Add up the time intervals sm to get time "t_j" in (3.9);

Step4. Construct the right side of the equation (3.9) to obtain time variable "t";

Step5. Calculate the integrand in (4.17) at the scaled time t, and change the sign of the integral. (This problem is seeking for a maximum. Since the computer package only deals with minimization calculation, an opposite sign of the objective function needs to be changed.)

Computational method 4.5 End-point condition (see model1_5.m in Appendix A.3)

Step1. Initialization. Input the last value of the first state function in vector xm and the parameters par which are required by the "End-point condition";

Step2. Construct the terminal term in (4.14).

7. Computing Results and Conclusion

In this section, the algorithms 4.1-4.5 are used in a computer software package (see in Appendix A.3) which was developed for this financial decision-making model (4.1)-(4.5). Before we present the computing results, the analytical solution in the Davis and Elzinga [22, 1970] finance model is described first for the future comparisons. Figure 6 in Davis and Elzinga [22, 1970] shows the optimum solution graphically, which means:

Solution case [1]: when $\frac{P(T)}{E(T)} > \frac{cr + \frac{k}{1-\delta}}{c\rho + k}$, the optimal solution has control in case [C] all the time;

Solution case [2]: when $\frac{P(T)}{E(T)} < \frac{cr + \frac{k}{1-\delta}}{c\rho + k}$, the optimal solution has control switching from case [B] to [A] at a switching time.

The control regions [A], [B], [C] are described in Section 4.3.

First, set the parameters:

$$\rho = 0.1, k = 0.15, r = 0.2, \delta = 0.1, c = 1$$

which meets the restriction on r/ρ in case (a) in Section 4.3.

$$\frac{1}{1-\delta} = \frac{1}{0.9} < r/\rho = 2 < \frac{c + [c + (1-\delta) + 1](k/\rho)}{c(1-\delta)} = 3.85/0.9$$

Initialize the states:

$$P(0) = 3, E(0) = 2$$

An Optimal Corporate Financing Model

Map the control (u_r, u_s) in this order:

[B] $u_r = k/r, u_s = 0$
[C] $u_r = 0, u_s = k/r$
[A] $u_r = 0, u_s = 0$

in successive time intervals. [B], [C], [A] for $nb = 3$, or [B], [C], [A], [B], [C], [A] for $nb = 6$.

Then run the programs for the algorithms 4.1-4.5 (details see "model1_1.m", "model1_2.m", "model1_3.m", "model1_4.m", "model1_5.m" in Appendix A.3). All the cases are put into Table 4.1.

Table 4.1. Computing results for solution case [1]

Case	T	nb	ns	-f	$\frac{P(T)}{E(T)}$	Control regions and switching times
Case 1	10	3	1	4.46	1.49	most [C]- switch at t=0.97 to [A]
Case 2	10	3	3	4.49	1.49	most [C]- switch at t=0.95 to [A]
Case 3	10	6	3	4.52	1.485	most [C]- [C] - switch at t=0.97 to [A]

Here, "most [C]" indicates that most of the time is spent in [C].

From the results in Table 4.1:

$$\frac{P(T)}{E(T)} = 1.49 > \frac{cr + \frac{k}{1-\delta}}{c\rho + k} = \frac{1*0.2 + \frac{0.15}{1-0.1}}{1*0.1 + 0.15} = 1.47$$

Then according the analytical solution in Davis' work, the optimal solution is supposed to be in solution case [1], which is control [C]. The computing results in Table 4.1 almost agree with the analytical solution. Since 1.49 is very close to 1.47, the case [A] will mix with case [C] at some point. This computed case is close to the theoretical optimum, but it does not exactly agree with it. Another solution case is verified next.

Initialize the states:

$$P(0) = 0.5, E(0) = 1$$

OPTIMAL CONTROL MODELS IN FINANCE

Map the optimal control (u_r, u_s) in this order:

[C] $u_r = 0, u_s = k/r$
[B] $u_r = k/r, u_s = 0$
[A] $u_r = 0, u_s = 0$

in successive time intervals. [C], [B], [A] for $nb = 3$, or [C], [B], [A], [C], [B], [A] for $nb = 6$.

Then run the programs. The results are shown in Table 4.2.

Table 4.2. Computing results for solution case [2]

Case	T	nb	ns	-f	$\frac{P(T)}{E(T)}$	Control regions and switching times
Case 1	10	3	3	2.03	1.03	[B]- switch at t=0.441 to [A]
Case 2	10	6	3	2.03	1.03	[B]- switch at t=0.439 to [A]

From the result in Table 4.2:

$$\frac{P(T)}{E(T)} = 1.03 < \frac{cr + \frac{k}{1-\delta}}{c\rho + k} = \frac{1*0.2 + \frac{0.15}{1-0.1}}{1*0.1 + 0.15} = 1.47$$

The computing results agree with the analytical solution in solution case [2], which is control case [B] switching to [A] at a certain switching time. In this example, the computed results agree well with the theory. Another example with the same initialization but a different given order of control mapping is given in Table 4.3; the results agree with Table 4.2.

The optimal control (u_r, u_s) are in order:

[B] $u_r = k/r, u_s = 0$
[C] $u_r = 0, u_s = k/r$
[A] $u_r = 0, u_s = 0$

Table 4.3. Computing results for solution case 2 with another mapping control

Case	T	nb	ns	-f	$\frac{P(T)}{E(T)}$	Control regions and switching times
Case 1	10	3	3	2.03	1.03	[B]- switch at t=0.441 to [A]
Case 2	10	6	3	2.03	1.03	[B]- switch at=0.441 to [A] -switch at 0.738 to [A]

An Optimal Corporate Financing Model

An approximate calculation using the SCOM package, dividing $[0, 1]$ into 20 equal subintervals, also confirms the switching patterns for solution case [2]. The following computation has the same initialization of the parameters as the above computation. The initial states also take: $P(0) = 0.5, E(0) = 1$. The solution of the computation is shown as follows.

The optimal control takes values:

$$
\begin{aligned}
(u_r, u_s) = & (k/r, 0), (k/r, 0), (k/r, 0), (k/r, 0), (k/r, 0), \\
& (k/r, 0), (k/r, 0), (k/r, 0), (k/r, 0), \\
& (0, 0), (0, 0), (0, 0), (0, 0), (0, 0), (0, 0), (0, 0), \\
& (0, 0), (0, 0), (0, 0), (0, 0)
\end{aligned}
$$

The states are:

$P(T) = 1.99$
$E(T) = 1.96$

The objective function is:

$-f = 2.04$

$P(T)/E(T) = 1.03 < \frac{cr + \frac{k}{1-\delta}}{c\rho + k} = \frac{1*0.2 + \frac{0.15}{1-0.1}}{1*0.1 + 0.15} = 1.47$, the optimal solution is expected to be in solution case [2] in the analytical solution. The results of this computation are very close to Table 4.2 and Table 4.3 and also confirm the analytical solution in Davis and Elzing's finance model. The optimal control jumps from [B] to [C] at $t = 0.45$.

The computation in this chapter agrees with the analytical solution in Davis and Elzinga [22, 1970]. The results might change if the parameters have been changed. The parameters set in this research are chosen to meet the restriction on r/ρ in case (a) (section 4.3). Further research on different parameters and more subdivisions of the time interval will be very interesting.

8. Optimal Financing Implications

The results of this computation in Tables 4.2 and 4.3 are very close to the analytical results in Section 3 and also confirm the analytical solution in Davis and Elzinga [22, 1970]. The model results provide the dynamic structure of capital of a firm and the optimal switching time from moving from one source of finance to another. There is one switch in the firm's financing strategy over the planning period. The levels of the two sources of fund depend on the relationships between the rate of return on equity capital (γ) and the investor's discount rate (ρ), and the relationship between the equity per share (E) and the market price of stock(P), thus the optimal control jumps from [B] to [C] at $t = 0.45$.

Since the computational results of the optimal financing model are consistent with the analytical results derived in Section 3 and the results of Davis and

Elzinga, they can be applied to understand optimal financing strategies of corporations to determine the optimal mix of structure of long-term funds to use in actual management of the capital structure of companies. Although theoretical controversies continue, model results suggest that the optimal structure which minimizes the firms' composite cost of capital changes over time. The model results provide the timing of switching from one fund to another fund. In real life, these switches depend on the cost of these funds, the rate of return, share prices, debt capacity, business cycles, business risks, etc. Various results are generated by different sets of parameter values of the model. Sometimes subjective judgments need to be made to choose the appropriate optimal capital mix path of the firm.

9. Conclusion

The determination of the optimal structure of corporate capital and the switching times for different methods of financing are essential for the actual management of capital structure of corporations. The computation of optimal switching time in this paper agrees with the analytical solution in Davis and Elzinga [22, 1970]. The results might change if the parameters are changed. The parameters set in this research is chosen to meet the restriction on r/ρ in case (a) (Section 4.3). Further research on different parameters and more subdivisions of the time interval will be very interesting. Development of an algorithm to coincide the time subdivisions with switching times are also another important area of further research.

Chapter 5

FURTHER COMPUTATIONAL EXPERIMENTS AND RESULTS

1. Introduction

This chapter will give some experiments that have been done for examining the computational algorithms developed in Chapter 2 and Chapter 3 for computing dynamic optimization financial models. The computation tests for the algorithms in Chapter 2 are presented in Section 5.2. Three computing examples are included. The problems met during computation are discussed and the solutions for those problems are indicated. Section 5.3 contains a different control policy with one possible pattern of the financial optimal control problem (3.2)-(3.6). The computing results of this new control policy are also included. The objectives of the modeling experiments are to provide examples of additional modeling structures which can be adopted in financial modeling including some new forms of the objective functions and control patterns, and to test the computational algorithms and the computer program by applying them to some further models.

2. Different Fitting Functions

In this section, three different given fitting functions of OCPWCS (Optimal Control Problems When Control are Step functions) are indicated. The difficulty and non-accuracy in the computation of financial models are analyzed and the possible solutions of those problems are given.

2.1 Calculation with square criterion in the objective function

Consider a financial optimal control model as follows:

$$MINJ = \int_0^1 (x(t) - 1/2t)^2 dt \qquad (5.1)$$

subject to:
$$\dot{x}(t) = u(t) \qquad (5.2)$$

$$u(t) = 1 \text{ or } 0 \qquad (5.3)$$

$$0 \leq t \leq 1 \qquad (5.4)$$

The target function is: 1/2*t.

Now apply the algorithms 2.1 - 2.4 to solve this problem. Note the objective function is no longer an absolute value. The algorithm 2.4 needs to be modified to meet the square value. The time division technique will be used here. The following graphs and tables are the computing results of this control problem. In the figures, "*-" represents the state function $x(t)$, and "." represents the given fitting function $1/2t$.

This model is the same investment planning model as in Section 2.7. However this model has a different fitting functions and the objective function is in a squared form which is a preferred form of decision criteria in economies (Islam [36, 2001]).

Figure 5.1 is the result of $n = 4$ (numbers of the time intervals).

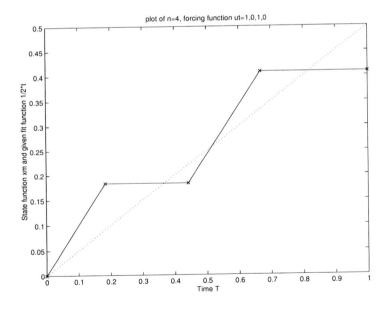

Figure 5.1. Plot of n=4, forcing function ut=1,0,1,0

Further Computational Experiments and Results

The outputs of this computation are shown as follows:
The real optimal switching times are:

$$v_1 = 0.183, v_2 = 0.442, v_3 = 0.667$$

Time period $[0, 1]$ is divided into four subintervals:

$$[0, 0.183], [0.183, 0.442], [0.442, 0.667], [0.667, 1]$$

The control is mapped as:

$$u(t) = 1, 0, 1, 0 \text{ in successive time intervals}$$

As results, the state variables of the financial system takes values as:

$$x(t) = 0.183, 0.183, 0.408, 0.408 \text{ at the ends of the time intervals}$$

The minimization of the objective function at $n = 4$ is:

$$J = 0.0028$$

Figure 5.2 is the graph result of $n = 10$.

112   OPTIMAL CONTROL MODELS IN FINANCE

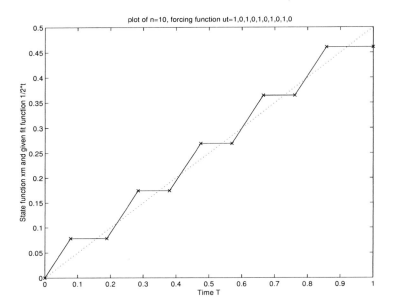

Figure 5.2.   Plot of n=10, forcing function ut=1,0,1,0,1,0,1,0,1,0

The outputs of $n = 10$ are:
The real optimal switching times are:

$$v_1 = 0.078, v_2 = 0.190, v_3 = 0.286, v_4 = 0.381,$$
$$v_5 = 0.476, v_6 = 0.572, v_7 = 0.668, v_8 = 0.763, v_9 = 0.858$$

Time period $[0, 1]$ is divided into 10 subintervals:

$$[0, 0.078], [0.078, 0.190], [0.190, 0.286], [0.286, 0.381], [0.381, 0.467],$$
$$[0.467, 0.572], [0.572, 0.668], [0.668, 0.763], [0.763, 0.858], [0.858, 1]$$

The control is mapped into:

$$u(t) = 1, 0, 1, 0, 1, 0, 1, 0, 1, 0 \text{ in successive time intervals}$$

As results, the state takes values as:

$$x(t) = \begin{array}{l} 0.078, 0.078, 0.174, 0.174, 0.270, 0.270, \\ 0.366, 0.366, 0.460, 0.460 \text{ at the ends of the time intervals} \end{array}$$

Further Computational Experiments and Results

The minimization of the objective function at $n = 10$ is:

$$J = 0.0005$$

When the jump of the control increases, a better approximation between state $x(t)$ and given fitting function is obtained.

In this experiment, several other computations have also been done in the case of $n = 2, 6$, and, 8. The results of the objective function of all these cases are shown in Figure 5.3 and in Table 5.1. The value of the objective function decreases when n increases. It is confirmed that the value of the objective function tends to zero when the control jumps infinitely often.

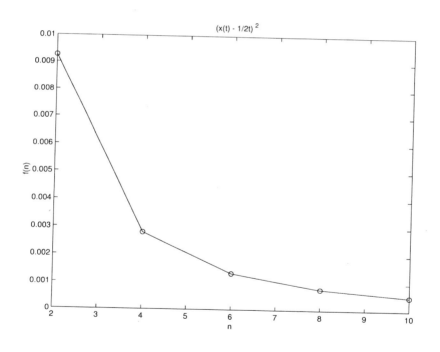

Figure 5.3. Results of objective function at n=2,4,6,8,10

The above results imply that for the development of a stable investment plan aimed at certain target value for the stock price, flexibility in switching among investment strategies is essential.

Table 5.1. Results of objective function at n=2,4,6,8,10

n	J
2	0.009259
4	0.002806
6	0.001335
8	0.000778
10	0.000508

2.2 Calculation with absolute value criterion in the objective function

In this section, a similar financial model similar experiments as in Section 5.2.1 are introduced. The difference is that the objective function in this section is the absolute value of the state approximating the given fitting function.

The financial optimal control problem is shown as follows:

$$MIN J = \int_0^1 |x(t) - 1/2t| dt \quad (5.5)$$

subject to:
$$\dot{x}(t) = u(t) \quad (5.6)$$

$$x_0 = 0 \quad (5.7)$$

$$u(t) = 1, \text{ or, } 0 \quad (5.8)$$

$$0 \leq t \leq 1 \quad (5.9)$$

The definition of the variables and parameters are same as in Section 3.2. However the present model has a different forcing function.

This is an optimal financing model for a firm where the decision problem involves whether to buy back (−1) or issue (1) some stocks or to maintain the present situation (0) of the amount of stocks used in the market so that the firm's stock price remains stable.

Apply the algorithms 2.1-2.4 on the problem (5.5)-(5.9). Only one modification is made in algorithm 2.4 for this given fitting function $1/2t$. There are two graphs shown as follows which represent $n = 4$ and $n = 8$ respectively. As in the last section, the lines represent "state function" by "*-", and "given fitting function" by ".".

The results gained from $n = 4$ are shown as follows:
The real optimal switching times are:

$$v_1 = 0.142, v_2 = 0.428, v_3 = 0.714$$

Time period $[0, 1]$ is divided into four subintervals:

$$[0, 0.142], [0.142, 0.428], [0.428, 0.714], [0.714, 1]$$

The control is mapped into:

$$u(t) = 1, 0, 1, 0 \text{ in successive time intervals}$$

As results, the state variable of the financial system variable of the financial system takes values as:

$$x(t) = 0.142, 0.142, 0.428 0.428 \text{ at the ends of the time intervals}$$

The minimization of the objective function at $n = 4$ is:

$$J = 0.032$$

Figure 5.4 also shows the approximation between the state function $x(t)$ and given fitting function $1/2t$. Three jumps of the optimal control are shown.

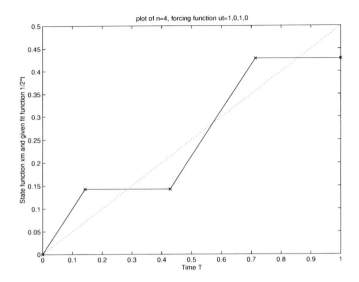

Figure 5.4. Plot of n=4, forcing function ut=1,0,1,0

An interesting situation was met during the computation at the case $n = 8$. Clearly the wrong result was gained in which the control only jumped once at $t = 0.333$ and the minimum of the objective function was $f = 0.064$, greater than the value of the minimum at the case $n = 4$. It is possible that the computation was finding some local minimum, instead of the global minimum. A check for this mistake was also made. It was found that error was in the initialization of the lengths of time intervals $um0$. Originally $um0$ was set as $[1/2, 1/2, 1/2, 1/2, 1/2, 1/2, 1/2]$. Since the constraint on time t is $0 \leq t \leq 1$, then the original set of $um0$ did not meet this restriction. It also misled the computation to a local minimum. The correct initialization of um is very critical for the whole computation. The mistake also occurred in the computation of the case $n = 8$ in Section 5.2.3. Now set $um0$ as $\frac{1}{nn}$. Accurate results of the problem are shown as follows.

The real optimal switching times are:

$$v_1 = 0.066, v_2 = 0.199, v_3 = 0.332, v_4 = 0.465,$$
$$v_5 = 0.598, v_6 = 0.731, v_7 = 0.865$$

Time period $[0, 1]$ is divided into four subintervals:

$[0, 0.066], [0.066, 0.199], [0.199, 0.332], [0.332, 0.465], [0.465, 0.598],$
$[0.598, 0.731], [0.731, 0.865], [0.865, 1]$

The control is mapped into:

$$u(t) = 1, 0, 1, 0, 1, 0, 1, 0 \text{ in successive time intervals}$$

The state takes values as:

$$\begin{aligned}x(t) =\ & 0.066, 0.066, 0.199, 0.199, 0.332, 0.332, \\ & 0.465, 0.466, 0.466 \text{ at the ends of the time intervals}\end{aligned}$$

The minimization of the objective function at $n = 8$ is:

$$J = 0.015$$

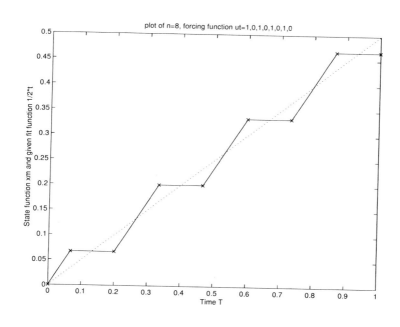

Figure 5.5. Plot of n=8, forcing function ut=1,0,1,0,1,0,1,0

Figure 5.5 also shows a better approximation than $n = 4$.

The results of other cases at $n = 2, 6, 10$ are shown in Table 5.2. The value of the objective function decreases when n (the number of the time intervals) increases.

Table 5.2. Results of objective functions at n=2,6,10

n	J
2	0.064818
4	0.032325
6	0.021356
8	0.015931
10	0.012599

2.3 Problem of the fitting function with big slope

In this section, a special case is introduced where the fitting function's slope is greater than 1 at time $t = 1$. Since the given fitting function is approximated by a piecewise-linear function with slope 1 and 0, the fitted function does not need any part of slope 0 near $t = 0.6$. In this case, the give fitting function has a slope 1.6 at $t = 1$, thus a slope 1 cannot give a good fit any more. A similar problem happened in Section 5.2.2 where the computation reached a local minimum also occurred at case $n = 8$.

Consider the following optimal control model of finance:

$$MINJ = \int_0^1 |x(t) - 0.8t^2| dt \quad (5.10)$$

subject to:

$$\dot{x}(t) = u(t) \quad (5.11)$$

$$x_0 = 0 \quad (5.12)$$

$$u(t) = 1 \text{ , or, } 0 \quad (5.13)$$

$$0 \leq t \leq 1 \quad (5.14)$$

After running the program of the algorithms 2.1-2.4, two mistakes were found. First, the control stopped jumping at case $n = 8$. Second, the approximation did not work well at the slope close to 1.

Further Computational Experiments and Results 119

The first problem was met while computing the optimal control problem (5.5)-(5.9). A different start of um can fix this problem. Here, we introduce several other methods to exam which one works well in the computation. First (method 1), simply divide each optimal um at case $n = 4$ by 2 as the initial start of um for case $n = 8$. The idea behind this method is that if the computation starts from a good minimum at the smaller jumps of control, it will also lead to a good minimum at the greater jumps of control. Second (method 2), since the non-accurate results always gave zero of um, which means no switching times, the lower bound of the um can be modified from the original zero to a very small number to avoid the situation of $um = 0$. Thus, the control jumps. It does not guarantee the computation will reach optimum. Sometimes these two methods can be combined together. Method 3, is the method used in Section 5.2.2 for solving a similar problem.

Using a control that can provide a bigger slope for the fitting function can solve the second problem mentioned earlier about the slope of the given fitting function. Method 4 lets control take a larger value to fit a function whose slope is great.

Figure 5.6 is the result of method 3. "*-" represents state function, "." represents the fitting function $0.8 * t^2$.

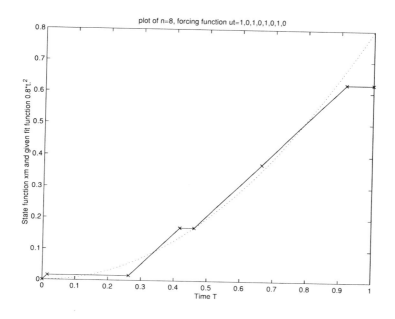

Figure 5.6. Plot of n=8, forcing function ut=1,0,1,0,1,0,1,0

The tests of all these methods are put in Table 5.3.

Table 5.3. Test results of the five methods

Case	lower bound	initial um	control	optimal switching times	J
1	ul=0.0	using method 1	1,0,...	0.019, 0.289, 0.502, 0.517, 0.692, 0.692, 0.924	0.039
2	ul=0.1	um0=1/2	1,0...	0.100, 0.297, 0.397, 0.497, 0.699, 0.799, 0.899	0.074
3	ul=0.05	um0=1/8	1,0,...	0.05, 0.25, 0.30, 0.35, 0.541, 0.601, 0.90, 0.99	0.051
4	ul=0.0	um0=1/8	1,0,...	0.015, 0.262, 0.416, 0.460, 0.662, 0.662, 0.917	0.038
5	ul=0.0	um0=1/8	1.5,0...	0.013, 0.307 0.422, 0.547, 0.932 0.718, 0.762, 0.932	0.028

J is the value of the objective function. These five cases represent four methods described earlier. Case 1 is for method 1, Case 2 is for method 2, Case 3 is for the combination of method 2 and method 3, Case 4 is for method 3, and Case 5 is for method 4. Method 1 and method 3 both obtain similar results. This proves that the right initialization of $um0$ will help the computation searching the optimum. Method 2 did not give promising results in the test. It can be explained that certain time intervals are not optimal, so if the control is forced to jump in that period, the optimum will be skipped. Method 5 does solve the problem with a big slope of the fitting function.

2.4 Conclusion

The three experiments that have been done in Section 5.2.1, 5.2.2, 5.2.3 are very important for the computer package which was developed for solving a certain class of optimal control problems. They make the program more general, the results more accurate, and computation more stable. The optimal control problems when controls are step functions can be solved using this computer package. Only very small modifications are needed. The accuracy gained from these computational algorithms is also very useful for the algorithms 3.1-3.4 for oscillator problems.

3. The Financial Oscillator Model when the Control Takes Three Values

In this section, an experiment when the control takes three values is introduced to test the algorithms 3.1-3.4 in Chapter 3. The results of the problem are also analyzed.

3.1 The control takes three values in an oscillator problem

Consider the financial oscillator model as follows:

$$MINJ = \int_0^1 |x_1(t) - 4*sin(5t+1)|dt \qquad (5.15)$$

subject to:

$$\dot{x}_1(t) = T*x_2(t) \qquad (5.16)$$
$$\dot{x}_2(t) = -T*x_1(t) + T*u(t) - T^2*B*x_2(t) \qquad (5.17)$$
$$u(t) \text{ takes value } -1, 0, 1, ... \text{ in successive time intervals} \qquad (5.18)$$
$$x_1(0) = 3 \qquad (5.19)$$
$$x_2(0) = 5 \qquad (5.20)$$
$$T = 5, B = 0.2 \qquad (5.21)$$

Apply the Algorithms 3.1-3.4 to this problem. There are two changes that need to be made in Algorithm 3.3 and Algorithm 3.4. One is re-mapping the control from original two-value pattern to three-value pattern. Here, the control is mapped to $-1, 0, 1$ in the successive time intervals. Note the smallest number of the subdivisions must be 3 because of the control values. Another change is in Algorithm 3.4 for the new fitting function $4*sin(5t+1)$.

Several cases are tried and the results are shown in Table 5.4. The time horizon $[0, 1]$ is first divided by $3, 6, 9$, thus becomes $3, 6, 9$ time intervals. Each time interval is further subdivided by $2, 3, 4, 6, 8$ to gain the better results of the integral in the objective function.

Table 5.4. Results of financial oscillator model

F	$n=3$	$n=6$	$n=9$
ns=1	1.0586	0.7707	1.5791
ns=2	1.1967	0.9828	0.9829
ns=3	0.9366	0.7030	0.5887
ns=4	0.9912	0.6361	0.7077
ns=6	0.7000	0.5586	0.5558
ns=8	0.6468	0.6441	0.5407

From the results in Table 5.4, a conclusion can be made that the subdivision of the time intervals is necessary especially when the nb (the number of the time intervals) is big. Note that in case $nb = 9$, $ns = 1$, the result of the objective function is 1.5791, nearly double of the result of $nb = 6$, $ns = 1$. This solution cannot be optimal. The results of states and optimal switching times in this case are shown as follows.

The real optimal switching times are:

$$v_1 = 0.316, v_2 = 0.488, v_3 = 0.488, v_4 = 1, v_5 = 1, v_6 = 1, v_7 = 1, v_8 = 1$$

The states take values:

$$xm_1 = [3\ 3\ 0.22\ 0.22\ -3.42\ -3.42\ -3.42\ -3.42\ -3.42\ -3.42]$$

$$xm_2 = [5\ -3.87\ -2.40\ -1.05\ -1.05\ -1.05\ -1.05\ -1.05\ -1.05\ -1.05]$$

The value of the objective function is:

$$J = 1.579$$

In this case, the control only jumps twice and then does not change. From the computation, we notice that a local minimum is reached instead of a global minimum. The increased number of subdivision of the subintervals will give more time for the system to reach the global minimum.

The chosen number for the subdivisions in the computation should be considered carefully. Since the control takes three values in this problem, the number of the big time intervals must be a multiplier of 3. Theoretically whether the number of small time intervals is a multiple of 3 does not matter for the computation. However, from the experimental results, we know that when the number of the subdivisions are the multiplier of 3, the computation gives a better approximation.

Figure 5.7 and Figure 5.8 are the graphical results in the case $nb = 9$ and $ns = 8$. In that case, the computation gives accurate results. Figure 5.7 shows the state function $x_1(t)$ and the fitting function $4*sin(5t+1)$. "o-" represents state function $x_1(t)$, "+:" represents the given fitting function $4*sin(5t+1)$.

Further Computational Experiments and Results 123

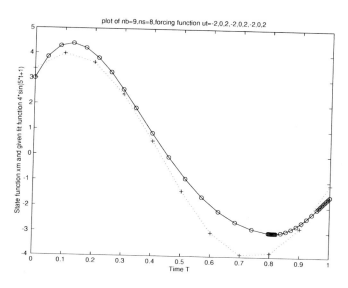

Figure 5.7. Plot of nb=9, ns=8, forcing function ut=-2,0,2,-2,0,2,-2,0,2

Figure 5.8 shows the relationship between two state functions during the time period [0, 1]. Since in this case, the state takes 72 values, we only give the graphical results here.

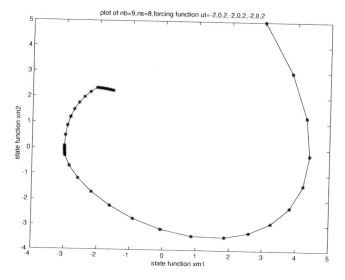

Figure 5.8. Relationship between two state functions during the time period 1,0

Graphic results of this oscillator problem are shown below. There are three sets of them representing the solutions at $nb = 3, 6, 9$.

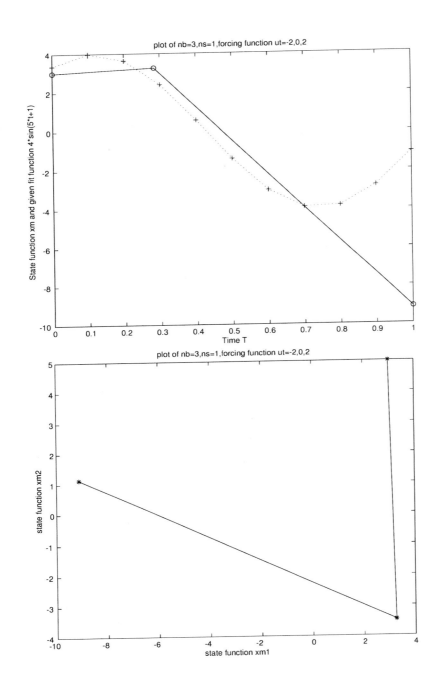

Further Computational Experiments and Results 125

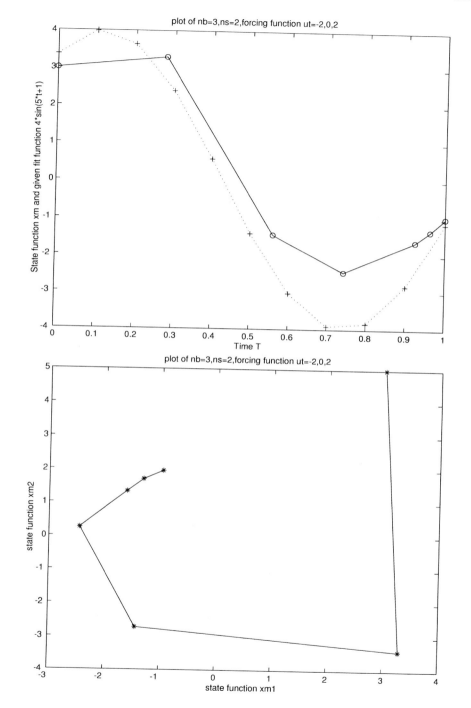

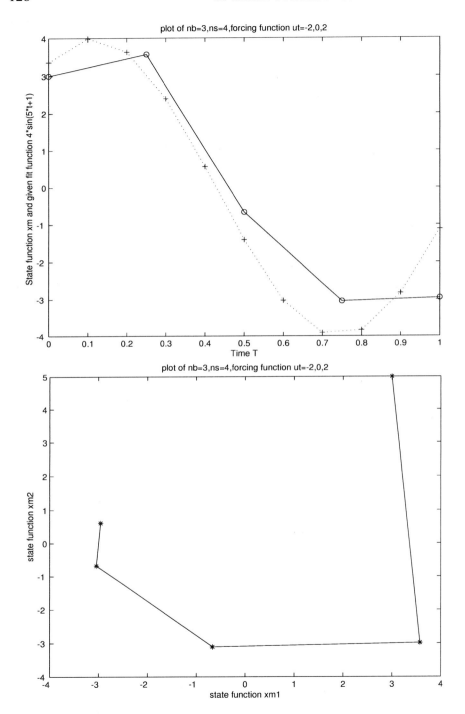

Further Computational Experiments and Results

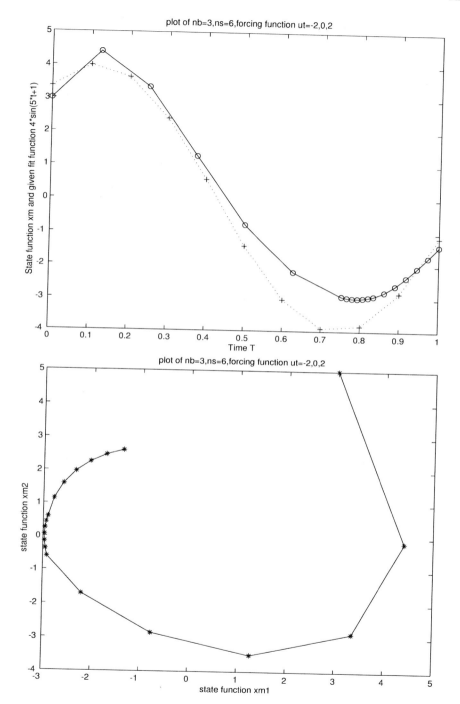

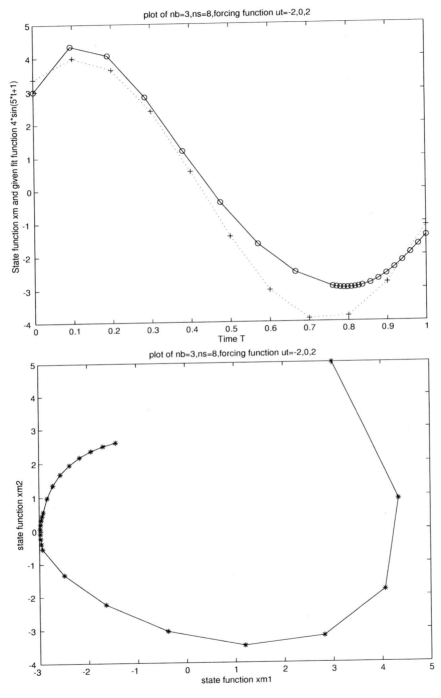

Set 1 for $nb = 3$

Further Computational Experiments and Results 129

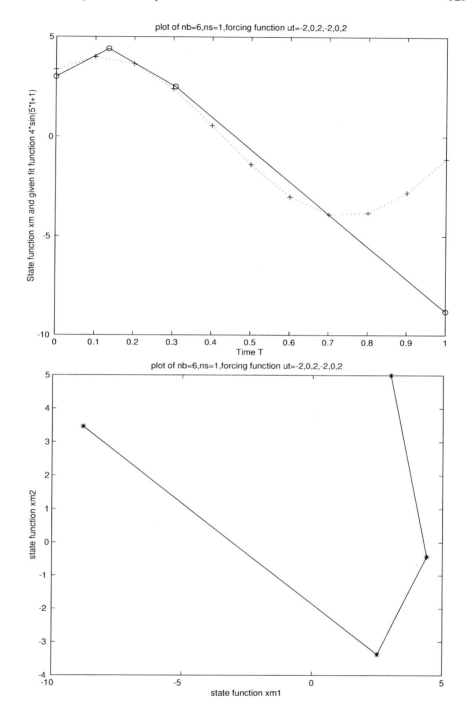

130 OPTIMAL CONTROL MODELS IN FINANCE

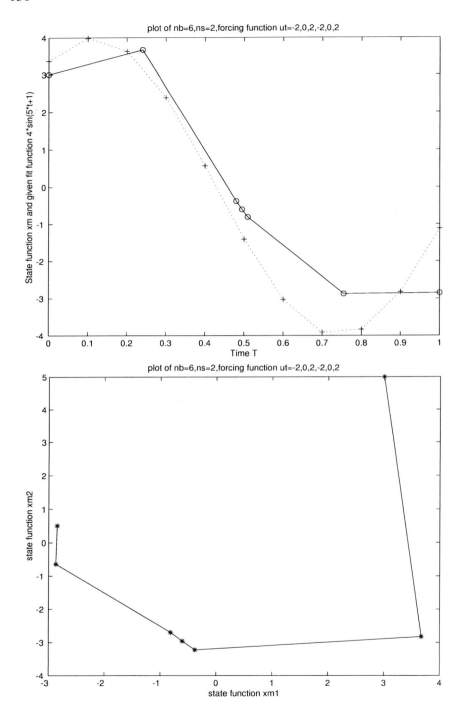

Further Computational Experiments and Results

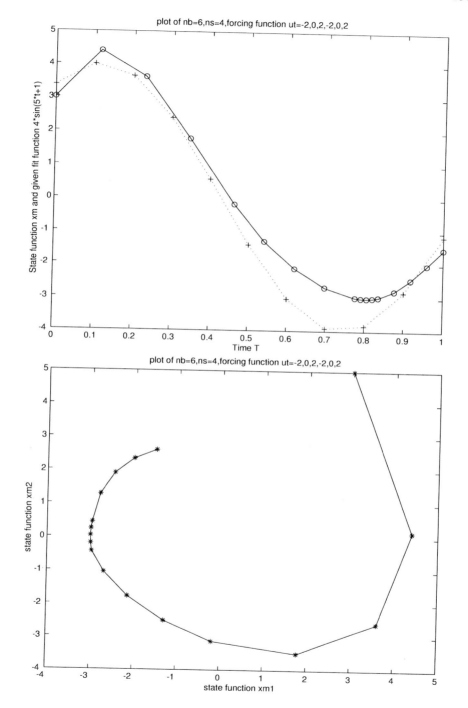

132 OPTIMAL CONTROL MODELS IN FINANCE

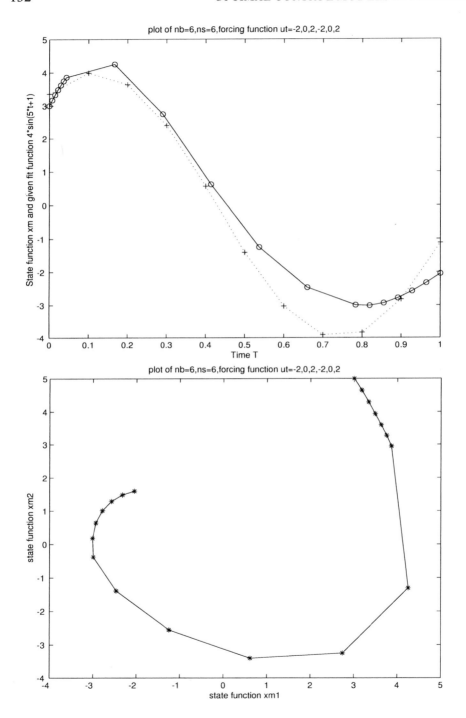

Further Computational Experiments and Results

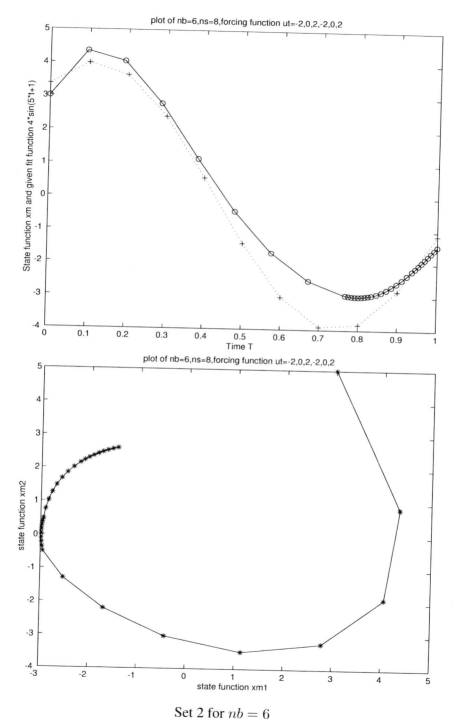

Set 2 for $nb = 6$

134 OPTIMAL CONTROL MODELS IN FINANCE

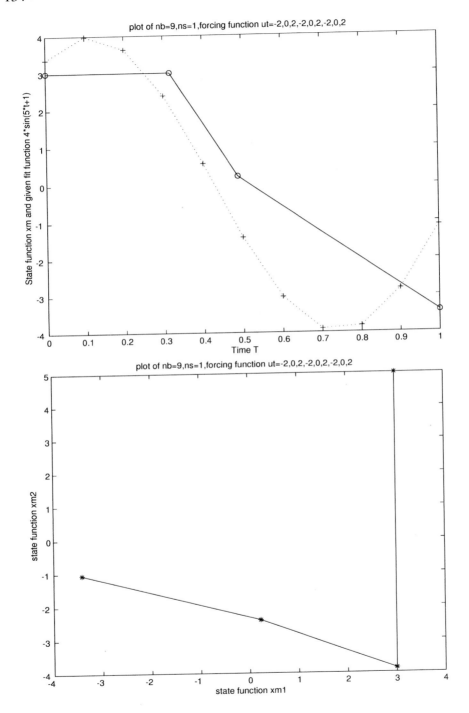

Further Computational Experiments and Results

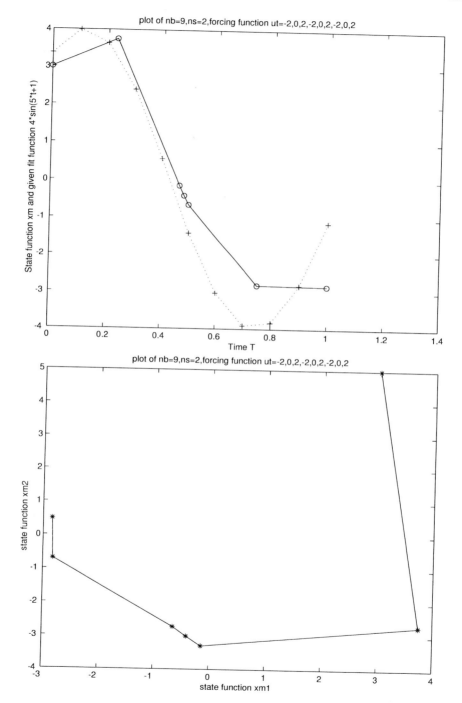

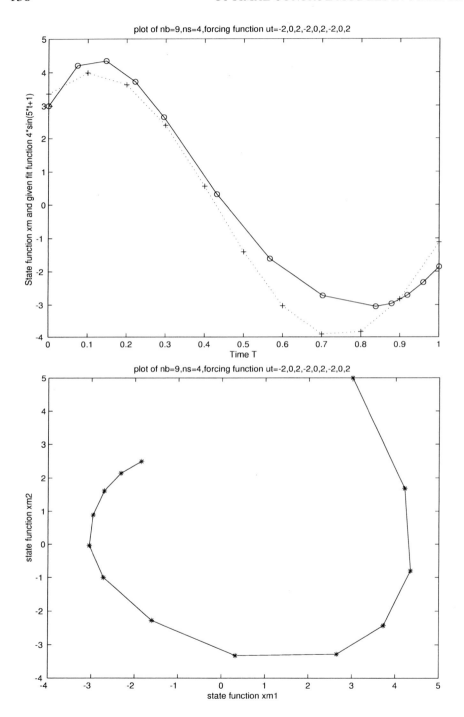

Further Computational Experiments and Results 137

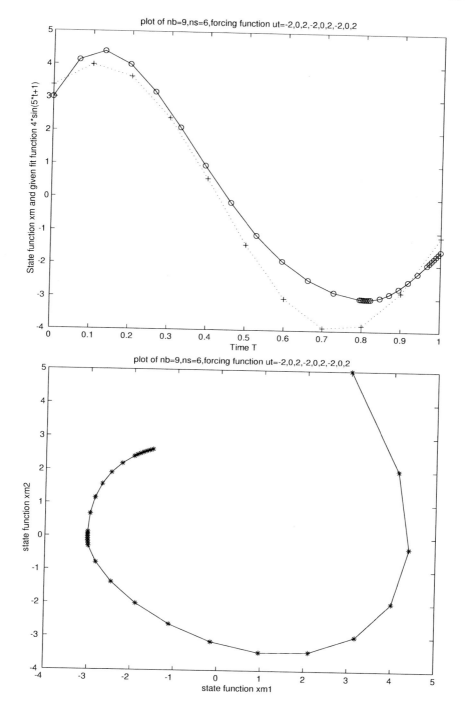

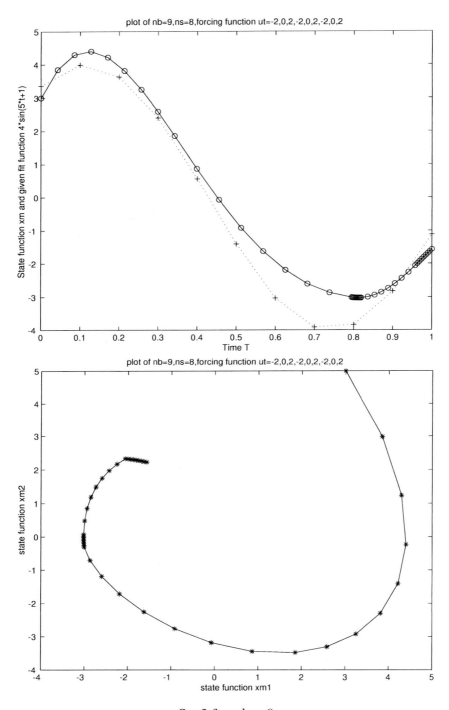

Set 3 for $nb = 9$

4. Conclusion

There are some financial optimal control models where the given fitting functions are *cos* or *sin* functions; then the financial control needs to take three values with the middle being zero, to ensure a good approximation. The experiment in Section 5.3.1 verifies the accuracy of the algorithms 3.1-3.4, thus the computer software package CSTVA (for details see Appendix A.4) based on these algorithms can solve all these kinds of control problems. The computed results provide insights into the dynamics of the financial system in terms of the state and control variables.

Chapter 6

CONCLUSION

Modeling and computation of dynamic optimization problems in finance is an important area for research in financial modeling. The thrust of this research has been to develop computational methods in order to solve financial optimal control models which are difficult to solve by traditional analysis using optimal control theories. Four computer software packages called CSTVA have been constructed, each of them used for different optimal control problems in two areas of finance: optimal investment planning and optimal corporate financing.

The STV approach consists of the following six major computational methods:

1. An optimization program based on the sequential quadratic programming (SQP)

2. The switching time variable method, the switching time is made a control variable.

3. The finite difference method for estimating gradients when gradients are not provided.

4. The step function approach to approximate the control variable.

5. A piecewise-linear (or non linear) transformation of time (as in MATLAB's "constr" program similar to the Newton Method for constrained optimization).

6. Second order differential equations represent the oscillatory dynamic financial models.

Financial optimal control modeling with a cost of changing control is the main topic in this book. Normally a cost is attached to a number of switching times then added to the objective function. The cost function becomes a new objective function to be treated. The chosen cost for such an optimal control problem for optimal investment planning in the stock market has been discussed in Chapter 2. The control in the problem is approximated by a step-function. The softwares constructed in this thesis compute the optimal switching times. Basically, the time period of the problem is divided into a certain number of subintervals to solve the differential equation and calculate the objective function as an integral. For the problem in Chapter 2, a greater number N of subdivisions will lead to a better fit to the target function $\phi(.)$. When a cost is attached to the switching times, the integral decreases as N increases, but the cost of switching increases with N. Hence the total cost function reaches an optimum. There are some techniques required for this research, such as the approximation of the control, the time scaled transformation for using the SCOM package "nqq" function, the piece-wise linear transformation for the calculation of the differential equation and integrals, non-linear transformation for the large time period, and penalty term transformations for the constraints of the problems. The computed results are also compared with the theoretical results. A certain class of optimal control problems can be put into the formula introduced in this research. The computer software packages developed in this research can then solve these problems with very little or no change.

Financing oscillator problems can form another class of financial optimal control models. These kinds of problems have a great number of applications in the real world. In these problems, the dynamic system can be described by a second-order differential equation. It is required to convert the second-order differential equation to two equivalent pair of first-order differential equations, to enable the software to be used. In order to get more accuracy in solving the differential equation, a further subdivision of the time intervals is introduced. While the control takes several discrete values, the sequences of these values may follow more than one pattern, leading to a different computed minimum. The computational algorithms are designed to handle these problems. The obtained computing results give good comparisons of them.

The modeling exercises in this book show the potential for modeling dynamic financial systems by adopting bang-bang optimal control methods. The results of this modeling can provide improved understanding about the behavior of and the decision problems in dynamic financial systems. The roles of switching times in financial strategies and transaction costs of controls are useful for financial planning as well.

The algorithms constructed in this thesis are applied to an optimal corporate financing model, which has two state functions and two control functions. The

Conclusion

computation experiments with two patterns of controls were tried. Effective results were obtained which also agree with the analytical solution from the original work. Another computation with the SCOM package also agreed with the computational solution in this research and the analytical solution. Further research of trying different parameters sets will be an interesting exercise. The accuracy of the computational algorithms in Chapter 3 and Chapter 4 were verified by these experiments. The STV approach may be considered satisfactory in view of its computational efficiency and time, and the plausibility of results.

In the damped oscillatory financial model, many corporate finance models are concerned with the application of optimal control. Computational approaches to the determination of the optimal financing problems were applied in a financial model which was first introduced by Davis and Elzinga [22, 1970]. The model discusses investment allocation in order to determine the optimal proportion of the sources of finance which can maximize the value of the company. In particular, this model determines the proportion of its earnings that should be retained for internal investment and what proportion should be distributed to shareholders as dividends. The model also aims to choose the "smart" investment program that gives the owners the most benefits.

In most of the existing optimal corporate financial structure models, the optimal proportion of various sources of funds is determined. In a linear dynamic finance model, this proportion may change depending on the bang-bang character of the time path in the model. For an improved understanding of the behavior of the dynamic path of such models, it is essential to know the optimal switching times for changing the optimal proportion of different funds. No algorithm exists in the literature which can determine the optimal switching time for corporate funds. The present book has developed such a model for optimal corporate financing and switching timing.

In this research, the computational algorithms have been improved for solving bang-bang optimal control problems. Applications of the STV algorithm to finance have been made to show the potential and methods for developing dynamic optimization methods in finance. Further research is necessary to improve the state of the art in computing bang-bang optimal control in general and financial optimal control models in particular.

Appendix A
CSTVA Program List

1. Program A: Investment Model in Chapter 2

```
Project1.m
function J=project1(C,nn)
J=project1_1(nn)+C*nn

Project1_1.m
% Program for the project 1 in chapter2

function project1_1=project1_1(nn)
figure;

par=[1,1,nn];
% parameters: 1 state, 1 control, nn subintervals.

options(14)=2000;
% maximum number of function evaluations.

options(13)=1;
% one equality constraint

xinit=0.0;
% initialize the state function xt

um0=ones(nn,1)/nn;
% take a initial guess of starting time intervals

s=[0:0.05:1];
plot(s,1/2*s,':')
ul=zeros(nn,1);
% lower bound of um, t(nn)-t(nn-1) >= 0

uu=ones(nn,1);
```

```
% upper bound of um, t(nn)-t(nn-1) <= 1

um=constr('project1_2',um0,options,ul,uu,[],xinit,par)
% use the MATLAB "constr" to get the optimal time intervals

[f,g,xm]=project1_2(um,xinit,par)
v(1)=um(1);
for ii=2:nn
        v(ii)=um(ii)+v(ii-1);
end
% obtain the real switching time t

v(nn)=0.9999999999999;
v=[0 v];
hold on
plot(v,xm,'rx-')
% plot the state function to switch time t

xlabel('Time T')
ylabel('State function xm and given fit function 1/2*t')


Project1_2.m
% Function of calculating differential equations and intervals

function [f,g,xm]=project1_2(um,xinit,par)
nx=par(1);
nu=par(2);
nn=par(3);
ps=1;
xm=zeros(nn+1,nx);
xm(1,:)=xinit;
lm=zeros(nn+1,nx);
% co-state function, polygonal function. In this case no
co-state function

ma=nx;
% Component of the state functions.

t=0;
% scale time t

it=1;
% counter it
hs=1/nn;
px='project1_3';
% form of the right side of the transformation of the
differential equation

xm=nqq(px,nx,nu,nn,xm,ma,t,it,hs,um,xm,lm,ps);
```

APPENDIX A: CSTVA Program List

```
% use SCOM package "nqq" to do the integral calculation

zz=zeros(nn+1,nx);
% here is the first xm in the calling function

zz(1,:)=0;
ma=1;
% set the number of the integral

t=0;
it=1;
hs=1/nn;
px='project1_4_1';
% form of the right side of the linear transformation of the
integral

jm=nqq(px,nx,nu,nn,zz,ma,t,it,hs,um,xm,lm,ps);
f=jm(nn+1);
% the result of the integral

g(1)=sum(um)-1;
% calculate the constraint


Project1_3.m
% Form of the right side of linear transformation of the
differential equation

function ff = project1_3(t,it,z,yin,hs,um,xm,lm,ps)
nn=1/hs;
rr=nn/2;
for i=1:rr,
        u(2*i-1)=1;
        u(2*i)=0;
end
% map the control pattern ut=0,1,... in successive time
intervals
pt=nn*um(floor(it),1);
ff=(u(1,floor(it)))*pt;
% Piecewise-linear transformation d/d(xt)=(1/h)*(t(j+1)-t(j))*u(j)


Project1_4_1.m
% Form of the right side of the integral

function ff = project1_4_1(t,it,z,yin,hs,um,xm,lm,ps)
nn=1/hs;
fr=nn*mod(t,hs);
% Linear interpolation
```

```
umx=zeros(nn,1);
umx(2) = um(1);
for ii=3:nn
        umx(ii)=umx(ii-1) + um(ii-1);
        end
xmt=(1-fr)*xm(it)+fr*xm(it+1);
% End-points of the time intervals

qt=nn*um(floor(it),1)*(t-hs*(it-1))+umx(floor(it),1);% time t
ll=abs (xmt-(1/2)*qt);
ff=nn*ll*um(floor(it),1); %|x(t)-1/2*t|
```

Project1_4_2.m
% A different case with the target function is (x(t)-1/2*t)^2

```
function ff = project1_4_2(t,it,z,yin,hs,um,xm,lm,ps)
nn=1/hs;
fr=nn*mod(t,hs);
umx=zeros(nn,1);
umx(2) = um(1);
for ii=3:nn
        umx(ii)=umx(ii-1) + um(ii-1);
        end
xmt=(1-fr)*xm(it)+fr*xm(it+1);
qt=nn*um(floor(it),1)*(t-hs*(it-1))+umx(floor(it),1);% time t
ll=(xmt-(1/2)*qt)^2;
ff=nn*ll*um(floor(it),1); %(x(t)-1/2*t)^2
```

Project1_4_3.m
% A different case with target is |x(t)-0.4*(t^2)|

```
function ff = project1_4_3(t,it,z,yin,hs,um,xm,lm,ps)
nn=1/hs;
fr=nn*mod(t,hs);
umx=zeros(nn,1);
umx(2) = um(1);
for ii=3:nn
        umx(ii)=umx(ii-1) + um(ii-1);
        end
xmt=(1-fr)*xm(it)+fr*xm(it+1);
qt=nn*um(floor(it),1)*(t-hs*(it-1))+umx(floor(it),1);
ll=abs (xmt-0.4*(qt^2));
ff=nn*ll*um(floor(it),1); % |x(t)-0.4*(t^2)|
```

project1_4_4.m
% a different case with target is |x(t)-0.8*(t^2)|

APPENDIX A: CSTVA Program List   149

```
function ff = project1_4_4(t,it,z,yin,hs,um,xm,lm,ps)
nn=1/hs;
fr=nn*mod(t,hs);
umx=zeros(nn,1);
umx(2) = um(1);
for ii=3:nn
        umx(ii)=umx(ii-1) + um(ii-1);
        end
xmt=(1-fr)*xm(it)+fr*xm(it+1);
qt=nn*um(floor(it),1)*(t-hs*(it-1))+umx(floor(it),1);
ll=abs (xmt-0.8*(qt^2));
ff=nn*ll*um(floor(it),1); % |x(t)-0.8*(t^2)|
```

2. Program B: Financial Oscillator Model in Chapter 3

```
Project2_1.m
%Program for project2, oscillator problem in chapter 3

function project2_1=project2_1(nb,ns) % nb is big interval of time
t and ns is small interval
par = [2,1,nb,ns];
xinit= [3,5];
%xinit=[6,5];
%par=[1,1,nb,ns];
%xinit=0.0;

nb=par(3); % big intervals
ns=par(4); % small intervals
nn=nb*ns; %total intervals
options(13)=1;
options(14)=10000;
um0=ones(nb,1)/nb;
%um0=[ 0.00168225289917 0.12189477200848 0.64070279483764
0.23572018024164];
%um0=[0.39023521914983 0.39023521914983 0.10976478085245
0.10976478085245];
%um0=[0 0.1390,0.5714,0.2896]';

ul=zeros(nb,1);
uu=ones(nb,1);
um=constr('project2_2',um0,options,ul,uu,[],xinit,par)
[f,g,xm] = project2_2(um,xinit,par)
sm=zeros(nn,1);
v=zeros(nn,1);
for ii =1: nb
        for jj= 1:ns
                sm((ii-1)*ns+jj)=um(ii)/ns;
        end
end
v(1)=sm(1);
```

```
for ii=2:nn
        v(ii)=sm(ii)+v(ii-1);
end
v(nn)=0.9999999999999;
s=zeros(nn+1,1);
s(1)=0;
for ii = 2: nn+1
        s(ii)=v(ii-1);
end
s
figure;
plot(s,xm(:,1),'ob-')
hold on;
t=[0:1/8   :1];
plot(t,-5*t+5, '+r:')
%plot(t,0.4*t.^2,'+r-')
%plot(t,t,'+r')

xlabel('Time T')
ylabel('State function xm and given fit function -5*t+t')
figure;
plot(xm(:,1),xm(:,2),'*b-')
xlabel('state function xm1')
ylabel('state function xm2')
%title('plot of switching time equal two')

%Project2_2.m integral
function [f,g,xm] = project2_2(um, xinit, par)
nx=par(1);
nu=par(2);
nb=par(3);
ns=par(4); % ns is small interval
nn=par(3)*par(4) ; % nn is total interval
sm=zeros(nn,1);
for ii =1: nb
        for jj= 1:ns
                sm((ii-1)*ns+jj)=um(ii)/ns;
        end
end
xm=zeros(nn+1,nx);
xm(1,:)=xinit;
lm=zeros(nn+1,nx);
ma=nx;
t=0;
it=1;
hs=1/nn;
px='project2_3';
xm=nqq(px,nx,nu,nn,xm,ma,t,it,hs,sm,xm,lm,ns);
zz=zeros(nn+1,nx);
zz(1,:) = 0;
```

APPENDIX A: CSTVA Program List

```
    ma=1;
    t=0;
    it=1;
    hs=1/nn;
    px='project2_4';
    x=xm(:,1);
    jm=nqq(px,nx,nu,nn,zz,ma,t,it,hs,sm,x,lm,ns);
    f=jm(nn+1);
    g(1)=sum(sm)-1;


    %Project2_3.m compute differential equation

    function ff = project2_3(t,it,z,yin,hs,sm,xm,lm,ns)
    nn=1/hs;
    T=5;
    B=0.0;
    B=0.1;
    B=0.2;
    nb=nn/ns;
    rr=nn/(2*ns);
    %for i=1:rr,
    %       ut(2*i-1)=2;
    %       ut(2*i)=-2;
    %end
    for i=1:rr,
            ut(2*i-1)=2;
            ut(2*i)=-2;
    end
    ts=zeros(nn,1);
    if ns > 1
    for ii= 1:nb
            for jj=1:ns
                    ts(ii)=ts(ii)+sm((ii-1)*ns+jj);
            end
    end
    else
            for ii= 1: nb
                    ts(ii)=sm(ii);
            end
    end
    pt=nb*ts((floor((it-1)/ns)+1),1);
    %ff(1) = T*z(2)*nn*sm(1,floor(it));
    %ff(2) = (-T*z(1)+T*ut(1,floor((it-1)/ns)+1)-T^2*B*z(2))
    *nn*sm(1,floor(it));
    ff(1)= T * z(2)*pt;
    ff(2)= (-T * z(1) + T *ut(1,(floor((it-1)/ns)+1)) -
    T^2 * B *z(2))*pt;
    %ff=ut(1,(floor((it-1)/ns)+1))*nn*sm(1,floor(it));
    %ff=ut(1,(floor((it-1)/ns)+1))*pt;
```

152   OPTIMAL CONTROL MODELS IN FINANCE

```
%project2_4.m

function ff = project2_4(t,it,z,yin,hs,sm,x,lm,ns)
nn=1/hs;
fr=nn*mod(t,hs);
xmt=(1-fr)*x(it)+fr*x(it+1);
nb=nn/ns;
umx=zeros(nn,1);
umx(2)=sm(1);
for ii = 3: nn
        umx(ii)=sm(ii-1)+umx(ii-1);
end
qt=nn*sm(floor(it),1)*(t-hs*(it-1))+umx(floor(it),1);
%qt=nb*ts((floor((it-1)/ns)+1),1)*(t-hs*ns*((
floor((it-1)/ns)+1)-1));
%qt=qt+umx((floor((it-1)/ns)+1),1);
ll=abs (xmt-((-5)*qt+5));% |x1(t)-((-5*t)+5)|
%ll= (xmt-((-5)*qt+5))^2;
%ll=abs(xmt-0.4*(qt^2));
%ll=(xmt-qt)^2;
ff=ll*nn*sm(floor(it),1);
%ff=ll*pt;

%Project2_4_test.m
function ff = project2_4_test(t,it,z,yin,hs,sm,x,lm,ps)
nn=1/hs;
fr=nn*mod(t,hs);
xmt=(1-fr)*x(it)+fr*x(it+1);
umx=zeros(nn,1);
umx(2) = sm(1);
for ii=3:nn
        umx(ii)=umx(ii-1) + sm(ii-1);
end
u=zeros(nn/ps,1);
ts=zeros(nn/ps,1);

if ps > 1
for ii= 1:nn/ps
        u(ii*ps-(ps-1))=sm(ii*ps-(ps-1));
        for jj=ii*ps-(ps-2):ii*ps
                u(jj)=u(jj-1)+sm(jj);
        end
        ts(ii)=u(jj);
end
else
        for ii = 1: nn/ps
                ts(ii)=sm(ii)
        end
```

APPENDIX A: CSTVA Program List   153

```
end

pt=(nn/ps)*ts((floor((it-1)/ps)+1),1);
qt=nn*sm(1,floor(it))*(t-hs*(it-1))+umx(floor(it),1);
%ll=abs (xmt-((-5)*qt+5));% |x1(t)-((-5*t)+5)|
ll= (xmt-((-5)*qt+5))^2;
ff=nn*ll*sm(1,floor(it))*pt;
```

3. Program C: Optimal Financing Model in Chapter 4

```
% An application of a model of investment of an utility.
%Model1_1.m

function model1_1=model1_1(nb,ns,parameters) % parameters
here is used to define all the parameters in the model
%parameters = [ p k r d c T]; % parameters here is used
to define all the parameters in the model
%parameters=[ 0.1 0.15 0.2 0.1 1 1];
par = [2,2,nb,ns,parameters];
%xinit= [P0,E0];
%xinit=[1 1];
%xinit = [0.5,0.5];
%xinit=[1.5 1.5];
xinit=[3 2];

nb=par(3); % big intervals
ns=par(4); % small intervals
nn=nb*ns; %total intervals
options(13)=1;
options(14)=10000;
um0=ones(nb,1)/nb;
%um=[0.1,0.3,0.6]';

ul=zeros(nb,1);
uu=ones(nb,1);
um=constr('model1_2',um0,options,ul,uu,[],xinit,par);
[f,g,xm] = model1_2(um,xinit,par);
sm=zeros(nn,1);
v=zeros(nn,1);

for ii =1: nb
        for jj= 1:ns
                sm((ii-1)*ns+jj)=um(ii)/ns;
        end
end

v(1)=sm(1);

for ii=2:nn
        v(ii)=sm(ii)+v(ii-1);
```

154   OPTIMAL CONTROL MODELS IN FINANCE

```
end

v(nn)=0.9999999999999;
s=zeros(nn+1,1);
s(1)=0;
for ii = 2: nn+1
        s(ii)=v(ii-1);
end

s
figure;
plot(xm(:,1),xm(:,2),'*b-')

xlabel('state function xm1')


%Model1_2.m

function [f,g,xm] = model1_2(um,xinit,par)
nx=par(1);
nu=par(2);
nb=par(3);
ns=par(4); % ns is small interval
nn=par(3)*par(4) ; % nn is total interval
sm=zeros(nn,1);
for ii =1: nb
        for jj= 1:ns
                sm((ii-1)*ns+jj)=um(ii)/ns;
        end
end
xm=zeros(nn+1,nx);
xm(1,:)=xinit;
lm=zeros(nn+1,nx);
ma=nx;
t=0;
it=1;
hs=1/nn;
px='model1_3';

xm=nqq(px,nx,nu,nn,xm,ma,t,it,hs,sm,xm,lm,par);
zz=zeros(nn+1,nx);
zz(1,:) = 0;
ma=1;
t=0;
it=1;
hs=1/nn;
x=xm(:,2);
px='model1_4';

jm=nqq(px,nx,nu,nn,zz,ma,t,it,hs,sm,x,lm,par);
```

APPENDIX A: CSTVA Program List

```
%f=jm(nn+1);
g(1)=sum(sm)-1;
xf=xm(nn+1,1);
ed='model1_5';
f=jm(nn+1)+feval(ed,xf,par);


%Model1_3.m compute differential equation

function ff = model1_3(t,it,z,yin,hs,sm,xm,lm,par)
nn=1/hs;
nb=par(3);
ns=par(4);
rr=nn/(3*ns);
p=par(5);
r=par(7);
c=par(9);
T=par(10);
%for i=1:rr,
%       us(3*i-2)=0;
%       ur(3*i-2)=0;
%       us(3*i-1)=0;
%       ur(3*i-1)=par(6)/par(7);
%       us(3*i)=par(6)/par(7);
%       ur(3*i)=0;
%end

for i=1:rr,
        us(3*i-2)=par(6)/par(7);   % [C]
        ur(3*i-2)=0;
        us(3*i-1)=0;               % [B]
        ur(3*i-1)=par(6)/par(7);
        us(3*i)=0;                 % [A]
        ur(3*i)=0;
end

ts=zeros(nn,1);

if ns > 1
for ii= 1:nb
        for jj=1:ns
                ts(ii)=ts(ii)+sm((ii-1)*ns+jj);
        end
end
else
        for ii= 1: nb
                ts(ii)=sm(ii);
        end
end
```

```
pt=nb*ts(floor((it-1)/ns)+1,1);
ff(1)= T*c*([1-ur(1,floor((it-1)/ns)+1)]*r*z(2)-p*z(1))*pt;
ff(2)= T*r*z(2)*[ur(1,floor((it-1)/ns)+1)+us(1,
floor((it-1)/ns)+1)*(1-z(2)/((1-d)
*z(1)))]*pt;

%Model1_4.m

function ff = model1_4(t,it,z,yin,hs,sm,xm,lm,par)
nn=1/hs;
fr=nn*mod(t,hs);
xmt=(1-fr)*xm(it)+fr*xm(it+1);
nn=1/hs;
rr=nn/(3*ns);
p=par(5);
r=par(7);
T=par(10);
%for i=1:rr,
%       ur(3*i-2)=0;
%       ur(3*i-1)=0;
%       ur(3*i)=par(6)/par(7);
%end

for i=1:rr,
      ur(3*i-2)=0;
      ur(3*i-1)=par(6)/par(7);
      ur(3*i)=0;
end

umx=zeros(nn,1);
umx(2) = sm(1);
for ii=3:nn
        umx(ii)=umx(ii-1) + sm(ii-1);
        end
qt=nn*sm(floor(it),1)*(t-hs*(it-1))+umx(floor(it),1);
ll=exp(-p*qt*T)*[1-ur(1,floor(it))]*r*xmt;
ff=-nn*ll*sm(floor(it),1)*T;


%Model1_5.m

function ff = model1_5(xf,par)
p=par(5);
T=par(10);
ff=-xf*exp(-p*T);
```

4. Program D: Three Value-Control Model in Chapter 5

```
%Project3_1.m computation for the integral
```

APPENDIX A: CSTVA Program List

```
function project3_1=project3_1(nb,ns) % nb is big
interval of time t and ns is small interval
par = [2,1,nb,ns];
xinit= [3,5];
nb=par(3); % big intervals
ns=par(4); % small intervals
nn=nb*ns; %total intervals
options(13)=1;
options(14)=10000;
um0=ones(nb,1)/nb;
ul=zeros(nb,1);
uu=ones(nb,1);
um=constr('project3_2',um0,options,ul,uu,[],xinit,par);

[f,g,xm] = project3_2(um,xinit,par)
sm=zeros(nn,1);
v=zeros(nn,1);

for ii =1: nb
        for jj= 1:ns
                sm((ii-1)*ns+jj)=um(ii)/ns;
        end
end
v(1)=sm(1);
for ii=2:nn
        v(ii)=sm(ii)+v(ii-1);
end

v(nn)=0.9999999999999;
s=zeros(nn+1,1);
s(1)=0;
for ii = 2: nn+1
        s(ii)=v(ii-1);
end

s
figure;
plot(s,xm(:,1),'ob-')
hold on;
t=[0:1/10:1];

plot(t,4*sin(5*t+1), '+r:')
xlabel('Time T')
ylabel('State function xm and given fit function
4*sin(5*t+1) ')
figure;

plot(xm(:,1),xm(:,2),'*b-')
xlabel('state function xm1')
```

ylabel('state function xm2')

%Project3_2.m integral

```
function [f,g,xm] = project3_2(um, xinit, par)
nx=par(1);
nu=par(2);
nb=par(3);
ns=par(4); % ns is small interval
nn=par(3)*par(4) ; % nn is total interval
sm=zeros(nn,1);

for ii =1: nb
        for jj= 1:ns
                sm((ii-1)*ns+jj)=um(ii)/ns;
        end
end

xm=zeros(nn+1,nx);
xm(1,:)=xinit;
lm=zeros(nn+1,nx);
ma=nx;
t=0;
it=1;
hs=1/nn;
px='project3_3';
xm=nqq(px,nx,nu,nn,xm,ma,t,it,hs,sm,xm,lm,ns);

zz=zeros(nn+1,nx);
zz(1,:) = 0;
ma=1;
t=0;
it=1;
hs=1/nn;
px='project3_4';
x=xm(:,1);

jm=nqq(px,nx,nu,nn,zz,ma,t,it,hs,sm,x,lm,ns);
f=jm(nn+1);
g(1)=sum(sm)-1;
```


%Project3_3.m compute differential equation

```
function ff = project3_3(t,it,z,yin,hs,sm,xm,lm,ns)
nn=1/hs;
T=5;
B=0.0;
B=0.1;
```

APPENDIX A: CSTVA Program List

```
B=0.2;
nb=nn/ns;
rr=nn/(3*ns);
%for i=1:rr,
%       ut(3*i-1)=-1;
%       ut(3*i)=0;
%       ut(3*1+1)=1;
%end

for i=1:rr,
        ut(3*i-2)=-2;
        ut(3*i-1)=0;
        ut(3*i)=2;
end
ts=zeros(nn,1);

if ns > 1
for ii= 1:nb
        for jj=1:ns
                ts(ii)=ts(ii)+sm((ii-1)*ns+jj);
        end
end
else
        for ii= 1: nb
                ts(ii)=sm(ii);
        end
end

pt=nb*ts((floor((it-1)/ns)+1),1);
ff(1)= T * z(2)*pt;
ff(2)= (-T * z(1) + T *ut(1,(floor((it-1)/ns)+1)) -
T^2 * B *z(2))*pt;

%Project3_4.m
function ff = project3_4(t,it,z,yin,hs,sm,x,lm,ns)
nn=1/hs;
fr=nn*mod(t,hs);
xmt=(1-fr)*x(it)+fr*x(it+1);
nb=nn/ns;
umx=zeros(nn,1);
umx(2)=sm(1);

for ii = 3: nn
        umx(ii)=sm(ii-1)+umx(ii-1);
end
qt=nn*sm(floor(it),1)*(t-hs*(it-1))+umx(floor(it),1);
ll=abs(xmt-4*sin(5*qt+1));% |x1(t)-cos(0.4*t+2.85)|
ff=ll*nn*sm(floor(it),1);
```

Appendix B
Some Computation Results

1. Results for Program A

```
um: the optimal time intervals.
f: the result of the objective funtion
g: the result of the constraints
s: optimal switching times
xm: the values of the state variable

project1_1(2)
um =
     0.1271
     0.8729
f =
     0.0627
g =
  -4.5940e-11
xm =
          0
     0.1271
     0.1271

project1_1(4)
um =
     0.0299
     0.4868
     0.2912
     0.1921
f =
     0.0291
g =
  -7.8249e-13
xm =
```

```
             0
        0.0299
        0.0299
        0.3211
        0.3211
project1_1(6)
um =
        0.0154
        0.3612
        0.1226
        0.1763
        0.2135
        0.1111
f =
        0.0184
g =
      -3.4402e-11
xm =
             0
        0.0154
        0.0154
        0.1380
        0.1380
        0.3515
        0.3515
project1_1(8)
um =
        0.0108
        0.3063
        0.0801
        0.1569
        0.1200
        0.0971
        0.1539
        0.0749
f =
        0.0134
g =
      -7.3581e-12
xm =
             0
        0.0108
        0.0108
        0.0908
        0.0908
        0.2109
        0.2109
        0.3647
        0.3647
```

APPENDIX B: *Some Computation Results*   163

2. Results for Program B

```
project2_1(2,1)
um =
     0.4167
     0.5833
 f =
     1.8227
 g =
   -6.1625e-11
xm =
     3.0000    5.0000
     5.8281   -2.5048
     0.6973  -11.1396
 s =
          0
     0.4167
     1.0000
project2_1(2,2)
um =
     0.6512
     0.3488
 f =
     0.9394
 g =
   -2.5978e-12
xm =
     3.0000    5.0000
     5.3714   -1.6031
     2.3736   -1.6846
     0.3249   -2.6175
    -1.6526   -1.7606
 s =
          0
     0.3256
     0.6512
     0.8256
     1.0000
project2_1(4,1)
um =
     0.0584
     0.1610
     0.5106
     0.2700
 f =
     0.6311
 g =
   -2.8808e-11
xm =
     3.0000    5.0000
```

```
         4.2091    3.3108
         4.3875   -2.2196
         0.8660   -1.3209
        -1.3328   -1.5506
s =
              0
         0.0584
         0.2194
         0.7300
         1.0000
project2_1(4,2)
um =
         0.0123
         0.1431
         0.5868
         0.2578
f =
         0.3388
g =
      7.0286e-11
xm =
         3.0000    5.0000
         3.1513    4.8156
         3.2968    4.6323
         4.3621    1.4484
         4.4301   -0.9324
         2.5494   -1.2942
         1.4915   -0.1476
         0.8567   -1.6082
        -0.3231   -1.9003
s =
              0
         0.0062
         0.0123
         0.0839
         0.1554
         0.4488
         0.7422
         0.8711
         1.0000

project2_1(4,4)
um =
         0.0024
         0.1349
         0.5626
         0.3001
f =
         0.2786
g =
```

APPENDIX B: Some Computation Results

```
      -5.7630e-11
  xm =
       3.0000    5.0000
       3.0147    4.9824
       3.0293    4.9648
       3.0438    4.9472
       3.0583    4.9296
       3.7520    3.3204
       4.1877    1.8743
       4.3942    0.6038
       4.4016   -0.4854
       3.7206   -1.2868
       2.7950   -1.2454
       2.0640   -0.7997
       1.6791   -0.3053
       1.3593   -1.3175
       0.7498   -1.8604
       0.0120   -2.0176
      -0.7284   -1.8917
   s =
            0
       0.0006
       0.0012
       0.0018
       0.0024
       0.0361
       0.0698
       0.1036
       0.1373
       0.2779
       0.4186
       0.5592
       0.6999
       0.7749
       0.8499
       0.9250
       1.0000
project2_1(6,1)
  um =
       0.3257
       0.1238
       0.1980
      -0.0000
       0.3526
      -0.0000
   f =
       0.5290
   g =
      -1.4863e-10
  xm =
```

```
       3.0000    5.0000
       5.3715   -1.6041
       3.5555   -3.8604
       0.9604   -1.3485
       0.9604   -1.3485
       0.8020    0.9757
       0.8020    0.9757
s =
            0
       0.3257
       0.4495
       0.6474
       0.6474
       1.0000
       1.0000
project2_1(6,2)
um =
      -0.0000
       0.1329
       0.5063
       0.0741
       0.1196
       0.1671
f =
       0.2615
g =
     -4.3578e-11
xm =
       3.0000    5.0000
       3.0000    5.0000
       3.0000    5.0000
       4.1425    1.9787
       4.3902   -0.3727
       3.0702   -1.3534
       1.8010   -0.5648
       1.6448   -1.0998
       1.4012   -1.5101
       1.0408   -0.9124
       0.8471   -0.3992
       0.5008   -1.1814
      -0.0789   -1.5304
s =
            0
      -0.0000
      -0.0000
       0.0665
       0.1329
       0.3861
       0.6392
       0.6763
```

0.7133
0.7731
0.8329
0.9165
1.0000

3. Results for Program C

Diferent controls in model1_1

p=0.1, k=0.15, r=0.2, d=0.1, c=1, T=10,

x0=(5,3)

control: [C] [B] [A]

results:

model1_1(3,3,parameters)
um =
   0.94900897415069
  -0.00000000000000
   0.05099102586178
f =
  -6.98762419479152
g =
   1.247046910179961e-11
xm =
   5.00000000000000   3.00000000000000
   5.40600401539454   3.47094537715380
   5.95896569319826   3.95174665662712
   6.63077320213855   4.46391793267020
   6.63077320213855   4.46391793267020
   6.63077320213855   4.46391793267020
   6.63077320213855   4.46391793267020
   6.66948645920598   4.46391793267020
   6.70754726725807   4.46391793267020
   6.74496662226235   4.46391793267020
s =
                   0
   0.31633632471690
   0.63267264943379
   0.94900897415069
   0.94900897415069
   0.94900897415069
   0.94900897415069
   0.96600598277128
   0.98300299139188
   0.99999999999990

168 OPTIMAL CONTROL MODELS IN FINANCE

control: [A] [B] [C]

results:

model1_1(3,3,parameters)
um =
                         0
                         0
        1.00000000000000
f =
  -6.98439125301710
g =
      -2.220446049250313e-16
xm =
        5.00000000000000   3.00000000000000
        5.00000000000000   3.00000000000000
        5.00000000000000   3.00000000000000
        5.00000000000000   3.00000000000000
        5.00000000000000   3.00000000000000
        5.00000000000000   3.00000000000000
        5.00000000000000   3.00000000000000
        5.43239375551053   3.49626657386023
        6.02582277381470   4.00489471104744
        6.74935621984978   4.55053034609536
s =
                         0
                         0
                         0
                         0
                         0
                         0
                         0
        0.33333333333333
        0.66666666666667
        0.99999999999990

p=0.1, k=0.15, r=0.2, d=0.1, c=1, T=10,

x0=(3,2)

control: [C] [B] [A]

model1_1(3,3,parameters)
um =
    -0.00000000000000
     0.43627214331862
     0.56372785682345
f =

APPENDIX B: Some Computation Results

```
           -4.34971604717521
  g =
           1.420692452569483e-10
  xm =
        3.00000000000000    2.00000000000000
        3.00000000000000    2.00000000000000
        3.00000000000000    2.00000000000000
        3.00000000000000    2.00000000000000
        2.74560664232866    2.48750405168379
        2.56260603757432    3.09383820357164
        2.45034521298337    3.84796753331929
        3.34895792223138    3.84796753331929
        4.09363087946859    3.84796753331929
        4.71073523102452    3.84796753331929
  s =
                        0
       -0.00000000000000
       -0.00000000000000
       -0.00000000000000
        0.14542404777287
        0.29084809554574
        0.43627214331862
        0.62418142892643
        0.81209071453425
        0.99999999999990

x0=(5,3)

model1_1(3,3,parameters)
  um =
      0.94900897415069
     -0.00000000000000
      0.05099102586178
  f =
     -6.98762419479152
  g =
           1.247046910179961e-11
  xm =
        5.00000000000000    3.00000000000000
        5.40600401539454    3.47094537715380
        5.95896569319826    3.95174665662712
        6.63077320213855    4.46391793267020
        6.63077320213855    4.46391793267020
        6.63077320213855    4.46391793267020
        6.63077320213855    4.46391793267020
        6.66948645920598    4.46391793267020
        6.70754726725807    4.46391793267020
        6.74496662226235    4.46391793267020
  s =
                        0
```

```
       0.31633632471690
       0.63267264943379
       0.94900897415069
       0.94900897415069
       0.94900897415069
       0.94900897415069
       0.96600598277128
       0.98300299139188
       0.99999999999990
```

p=0.1, k=0.15, r=0.2, d=0.1, c=1, T=10,

x0=(3,2)

control: [A] [B] [C]

model1_1(3,3,parameters)
um =
                          0
                          0
       1.00000000000000
f =
      -4.49775108882148
g =
          -2.220446049250313e-16
xm =
       3.00000000000000   2.00000000000000
       3.00000000000000   2.00000000000000
       3.00000000000000   2.00000000000000
       3.00000000000000   2.00000000000000
       3.00000000000000   2.00000000000000
       3.00000000000000   2.00000000000000
       3.00000000000000   2.00000000000000
       3.36336204900022   2.27071366163561
       3.78530547911419   2.56979963470193
       4.26744142249872   2.90415715247832
s =
                          0
                          0
                          0
                          0
                          0
                          0
                          0
       0.33333333333333
       0.66666666666667
       0.99999999999990

x0=(5, 3)

APPENDIX B: Some Computation Results

```
model1_1(3,3,parameters)
um =
                  0
                  0
   1.00000000000000
f =
  -6.98439125301710
g =
     -2.220446049250313e-16
xm =
      5.00000000000000        3.00000000000000
      5.00000000000000        3.00000000000000
      5.00000000000000        3.00000000000000
      5.00000000000000        3.00000000000000
      5.00000000000000        3.00000000000000
      5.00000000000000        3.00000000000000
      5.00000000000000        3.00000000000000
      5.43239375551053        3.49626657386023
      6.02582277381470        4.00489471104744
      6.74935621984978        4.55053034609536
s =
                  0
                  0
                  0
                  0
                  0
                  0
                  0
   0.33333333333333
   0.66666666666667
   0.99999999999990


p=0.1, k=0.15, r=0.2, d=0.1, c=1, T=10,

x0=(0.5, 1)

control :[C], [B], [A]

results:

model1_1(3,3,parameters)
um =
  -0.00000000000000
   0.44198331380348
   0.55801668619645
f =
  -2.03958198064794
g =
```

172   *OPTIMAL CONTROL MODELS IN FINANCE*

```
      -6.283862319378386e-14
xm =
   0.50000000000000   1.00000000000000
   0.50000000000000   1.00000000000000
   0.50000000000000   1.00000000000000
   0.50000000000000   1.00000000000000
   0.50836518053403   1.24730846467207
   0.53459240084601   1.55577840604261
   0.58093560157541   1.94053557501097
   1.14106504606018   1.94053557501097
   1.60612415480481   1.94053557501097
   1.99224914208392   1.94053557501097
s =
                  0
  -0.00000000000000
  -0.00000000000000
  -0.00000000000000
   0.14732777126783
   0.29465554253566
   0.44198331380348
   0.62798887586897
   0.81399443793445
   0.99999999999990


p=0.1, k=0.15, r=0.2, d=0.1, c=1, T=10,

x0=(0.5, 1)

control :[A], [B], [C]

results:

model1_1(3,3,parameters)
um =
   0.52071677552328
  -0.00000000000000
   0.47928322447289
f =
  -1.81807434266798
g =
    -3.836042594684841e-12
xm =
   0.50000000000000   1.00000000000000
   0.73901347970328   1.00000000000000
   0.93994199708665   1.00000000000000
   1.10885409839574   1.00000000000000
   1.10885409839574   1.00000000000000
   1.10885409839574   1.00000000000000
   1.10885409839574   1.00000000000000
```

APPENDIX B: Some Computation Results

```
     1.24170396001051    1.01227203398012
     1.36155842306688    1.04280425406198
     1.47466384342158    1.08536853674905
  s =
                       0
     0.17357225850776
     0.34714451701552
     0.52071677552328
     0.52071677552328
     0.52071677552328
     0.52071677552328
     0.68047785034757
     0.84023892517187
     0.99999999999990

model1_1(6,6,parameters)
 um =
     0.39890660392318
    -0.00000000000000
     0.00000000000000
     0.11928746513786
    -0.00000000000000
     0.48180593092858
 f =
    -1.82053330509933
 g =
     -1.038513719464618e-11
 xm =
     0.50000000000000    1.00000000000000
     0.59648376317515    1.00000000000000
     0.68676144864600    1.00000000000000
     0.77123224690021    1.00000000000000
     0.85026967149177    1.00000000000000
     0.92422321064582    1.00000000000000
     0.99341987262807    1.00000000000000
     0.99341987262807    1.00000000000000
     0.99341987262807    1.00000000000000
     0.99341987262807    1.00000000000000
     0.99341987262807    1.00000000000000
     0.99341987262807    1.00000000000000
     0.99341987262807    1.00000000000000
     0.99341987262807    1.00000000000000
     0.99341987262807    1.00000000000000
     0.99341987262807    1.00000000000000
     0.99341987262807    1.00000000000000
     0.99341987262807    1.00000000000000
     1.01323431734672    1.00000000000000
     1.03265871639605    1.00000000000000
     1.05170074779209    1.00000000000000
```

```
            1.07036793840976   1.00000000000000
            1.08866766695811   1.00000000000000
            1.10660716689690   1.00000000000000
            1.10660716689690   1.00000000000000
            1.10660716689690   1.00000000000000
            1.10660716689690   1.00000000000000
            1.10660716689690   1.00000000000000
            1.10660716689690   1.00000000000000
            1.10660716689690   1.00000000000000
            1.17569510221352   1.00307087089333
            1.24040740530816   1.01202132418871
            1.30187570955606   1.02549798919538
            1.36097381645976   1.04259929524034
            1.41839237440713   1.06270178723606
            1.47468755717307   1.08536187030624
s =
                         0
            0.06648443398720
            0.13296886797439
            0.19945330196159
            0.26593773594879
            0.33242216993598
            0.39890660392318
            0.39890660392318
            0.39890660392318
            0.39890660392318
            0.39890660392318
            0.39890660392318
            0.39890660392318
            0.39890660392318
            0.39890660392318
            0.39890660392318
            0.39890660392318
            0.39890660392318
            0.41878784811282
            0.43866909230247
            0.45855033649211
            0.47843158068175
            0.49831282487140
            0.51819406906104
            0.51819406906104
            0.51819406906104
            0.51819406906104
            0.51819406906104
            0.51819406906104
            0.51819406906104
            0.59849505754914
            0.67879604603723
            0.75909703452533
```

APPENDIX B: Some Computation Results   175

```
    0.83939802301342
    0.91969901150152
    0.99999999999990
```

4. Results for Program D

```
project3_1(3,1)
f =
    1.0586
g =
-6.2238e-11
xm =
    3.0000    5.0000
    3.2515   -3.4665
    3.2515   -3.4665
   -9.1222    1.1345
s =
         0
    0.2840
    0.2840
    1.0000

project3_1(3,2)
f =
    1.1967
g =
-1.4602e-10
xm =
    3.0000    5.0000
    3.2944   -3.4031
   -1.4369   -2.7269
   -2.4523    0.2524
   -1.6076    1.3348
   -1.3167    1.7035
   -0.9671    1.9524
s =
         0
    0.2792
    0.5584
    0.7411
    0.9239
    0.9620
    1.0000

project3_1(3,3)
f =
    0.9367
g =
  1.7265e-10
```

176 OPTIMAL CONTROL MODELS IN FINANCE

```
xm =
    3.0000    5.0000
    3.7672   -2.6588
   -0.0810   -3.3194
   -2.7291   -1.1775
   -2.8827    0.4586
   -2.3886    1.3429
   -1.6078    1.5983
   -1.6078    1.5983
   -1.6078    1.5983
   -1.6078    1.5983
s =
         0
    0.2302
    0.4604
    0.6906
    0.7938
    0.8969
    1.0000
    1.0000
    1.0000
    1.0000

project3_1(6,1)
f =
    0.7708
g =
    2.0042e-10
xm =
    3.0000    5.0000
    4.3970   -0.4306
    4.3970   -0.4306
    4.3970   -0.4306
    2.5059   -3.3710
    2.5059   -3.3710
   -8.8124    3.4589
s =
         0
    0.1346
    0.1346
    0.1346
    0.3056
    0.3056
    1.0000

project3_1(6,2)
f =
    0.9828
g =
-1.9679e-10
```

APPENDIX B: Some Computation Results

```
xm =
    3.0000    5.0000
    3.6722   -2.8237
   -0.3772   -3.2174
   -0.6028   -2.9559
   -0.8093   -2.6976
   -0.8093   -2.6976
   -0.8093   -2.6976
   -2.8731   -0.6405
   -2.8489    0.5052
   -2.8489    0.5052
   -2.8489    0.5052
   -2.8489    0.5052
   -2.8489    0.5052
s =
         0
    0.2400
    0.4800
    0.4946
    0.5092
    0.5092
    0.5092
    0.7546
    1.0000
    1.0000
    1.0000
    1.0000
    1.0000

project3_1(9,1)
f =
    1.5792
g =
-1.1642e-10
xm =
    3.0000    5.0000
    3.0007   -3.8736
    0.2249   -2.4065
    0.2249   -2.4065
   -3.4249   -1.0539
   -3.4249   -1.0539
   -3.4249   -1.0539
   -3.4249   -1.0539
   -3.4249   -1.0539
   -3.4249   -1.0539
s =
         0
    0.3164
    0.4882
    0.4882
```

```
      1.0000
      1.0000
      1.0000
      1.0000
      1.0000
      1.0000

project3_1(9,2)
f =
    0.9830
g =
  2.2104e-10
xm =
    3.0000    5.0000
    3.7478   -2.6935
   -0.1429   -3.2994
   -0.4100   -3.0087
   -0.6527   -2.7209
   -0.6527   -2.7209
   -0.6527   -2.7209
   -2.8224   -0.6915
   -2.8407    0.4891
   -2.8407    0.4891
   -2.8407    0.4891
   -2.8407    0.4891
   -2.8407    0.4891
   -2.8407    0.4891
   -2.8407    0.4891
   -2.8407    0.4891
   -2.8407    0.4891
   -2.8407    0.4891
   -2.8407    0.4891
s =
         0
    0.2322
    0.4644
    0.4814
    0.4983
    0.4983
    0.4983
    0.7492
    1.0000
    1.0000
    1.0000
    1.0000
    1.0000
    1.0000
    1.0000
    1.0000
    1.0000
    1.0000
```

APPENDIX B: Some Computation Results

```
1.0000
1.0000
```

Appendix C
Differential Equation Solver from the SCOM Package

```
function  xk=nqq(pd,nx,nu,nn,fil,ma,t,it,hs,um,xm,lm,ps)
% Calling the Runge-Kutta integration

yin=fil(1,:);xk(1,:)=yin;
while it < nn+1
[y2,it2,t2]=rqq(pd,ma,t,it,hs,yin,nx,nu,nn,um,xm,lm,ps);
xk(it+1,:)=y2;
it=it2;t=t2;
yin=y2;
end


function [y1,it,t]=rqq(pd,ma,t,it,hs,yin,nx,nu,nn,um,xm,lm,ps)
% Runge-Kutta stages

fp=zeros(ma,1)'; tt=zeros(ma,1)';
p=0; q=1;
tz=kqq(pd,ma,t,it,hs,p,q,fp,yin,tt,um,xm,lm,ps);
tt=tz(1,:); fp=tz(2,:);
p=0.5;q=2;t=t+0.5*hs;
tz=kqq(pd,ma,t,it,hs,p,q,fp,yin,tt,um,xm,lm,ps);
tt=tz(1,:); fp=tz(2,:);
tz=kqq(pd,ma,t,it,hs,p,q,fp,yin,tt,um,xm,lm,ps);
tt=tz(1,:); fp=tz(2,:);
t=t+0.5*hs;p=1;q=1;
tz=kqq(pd,ma,t,it,hs,p,q,fp,yin,tt,um,xm,lm,ps);
tt=tz(1,:); fp=tz(2,:);
it=it+sign(hs);
it;
y1=yin+tt/6;
```

```
function tz=kqq(pd,ma,t,it,hs,p,q,fp,yin,tt,um,xm,lm,ps)
% Calling the function

z=yin+p*hs*fp;
ff=feval(pd,t,it,z,yin,hs,um,xm,lm,ps);
fp=ff;
tz=[tt+q*hs*fp;fp] ;


function ix=jqq(vv,hh,xi)
% Linear interpolation

global um xm jm
nn=length(vv')-1;
nz=-1/hh;
if xi < 0, xi = 0; end
fr=-nz*mod(xi,hh);
bb=ceil(xi*nz)+1;

if bb>1, vw=vv(bb-1,:);
else vw=vv(1);
end

ix=(1-fr)*vv(bb,:) + fr*vw;
```

Appendix D
SCOM Package

```
subs=cell(1,9);
subs={'dnx','dnj','dnf','dnc'};

par=[2, 2, 20, 0, 0, 1, 6, 0.2, 0.1, 0.1, 0.75];
% nx, nu, nn, npa, grad, c, T, r, p, d , b=k/r\
% nx = number of states, nu = number of controls,
  nn = number of subdivisions
% npa = 0 (reserved), grad = 1 if gradients calculated,
otherwise 0
% The remaining parameters, specific to the problem,
are passed to the subroutines

xinit=[2 1]; nn=par(3); % Initial values for the state(s)
u0=zeros(nn,2); % Starting values for computing the control
ul=zeros(nn,2); % Lower bound (vector or matrix) for the control
uu=ones(nn,2); % Upper bound (vector or matrix) for the control

figure
Control=constr('fqq',u0,[],ul,uu,[],par,subs,xinit)
% Calls constr package

[Objective,Constraint,State,Integral]= cqq(Control,par,subs,xinit)
% Computes optimal state, etc., from the optimal control
  got from constr

% If gradients are calculated, then used instead
% [Objective,Constraint,State,Integral,Costate,Gradient]
  =cqq(Control,par,subs,xinit)
% The following lines plot one state component against another,
  and plot two control components, ehich are step-functions,
  against time
```

```
plot(State(:,1),State(:,2),'x-')
xlabel('State 1')
ylabel('State 2')

figure
t2=[0:0.001:0.999];
plot(t2,Control(floor(nn*t2)+1,1),'r')

hold on
plot(t2,Control(floor(nn*t2)+1,2),'g')
%The following codes comprise the SCOM package;
 the user does not alter them

% Get function values as required by constr


function [f,g]=fqq(uz,ps,q,xinit) % Input data to the subroutine

if ps(5)==1
[a1,a2,a3,a4,a5,a6]=cqq(uz,ps,q,xinit);
% Output data
elseif ps(5)==0
[a1,a2,a3,a4]=cqq(uz,ps,q,xinit);
end

f=a1;
g=a2;


function [df,dg]=gqq(uu,ps,q,xinit)
% Get gradient values as required by constr

[a1,a2,a3,a4,a5,a6]=cqq(uu,ps,q,xinit);
df=a6';
nn=ps(3);
pk=char(q(5));
dg=feval(pk,uu,nn);


function [f,g,xm,f0,lm,gr]=cqq(um,ps,q,xinit)
% Solve differential equations

nx=ps(1);nu=ps(2);nn=ps(3);npa=ps(4); rec=0;npar=1;
xm=zeros(nn+1,nx); xm(1,:)=xinit; lm=zeros(nn+1,nx);
ma=nx;t=0;it=1;hs=1/nn;
px=char(q(1));
xm=nqq(px,nx,nu,nn,xm,ma,t,it,hs,um,xm,lm,ps); %compute state
ma=1;t=0;it=1;hs=1/nn;
zz=zeros(nn+1,1); zz(1,:)=0;
```

APPENDIX D: SCOM Package

```
pj=char(q(2));

jm=nqq(pj,nx,nu,nn,zz,ma,t,it,hs,um,xm,lm,ps);%compute integral
xf=xm(nn+1,:) ;
pf=char(q(3));
f=jm(nn+1) + feval(pf,xf,um,xm,ps); %objective
pc=char(q(4));

for ii=1:nn
g(ii)=feval(pc,ii,hs,um,xm,lm,ps); %control constraint
end

f0=jm(nn+1);
%final state

if ps(5)==1
%if gradients are supplied

ma=nx;t=1;it=nn;hs=-1/nn;
lm=zeros(nn+1,nx);
pa=char(q(8));
lm(nn+1,:)=feval(pa,nn,xf,um,xm,ps);
pq=char(q(6));
lm=lqq(pq,nx,nu,nn,lm,ma,t,it,hs,um,xm,lm,ps);
%costate


function xk=nqq(pd,nx,nu,nn,fil,ma,t,it,hs,um,xm,lm,ps)
%Runge-Kutta (organize steps for fourth order RK
  integration of DEs)

yin=fil(1,:);xk(1,:)=yin;

while it < nn+1
[y2,it2,t2]=rqq(pd,ma,t,it,hs,yin,nx,nu,nn,um,xm,lm,ps);
xk(it+1,:)=y2;
it=it2;t=t2;
yin=y2;
end

function [y1,it,t]=rqq(pd,ma,t,it,hs,yin,nx,nu,nn,um,xm,lm,ps)
% Runge-Kutta : increments

fp=zeros(ma,1)'; tt=zeros(ma,1)';
p=0; q=1;
tz=kqq(pd,ma,t,it,hs,p,q,fp,yin,tt,um,xm,lm,ps);
tt=tz(1,:); fp=tz(2,:);
p=0.5;q=2;t=t+0.5*hs;

tz=kqq(pd,ma,t,it,hs,p,q,fp,yin,tt,um,xm,lm,ps);
```

```
tt=tz(1,:); fp=tz(2,:);
tz=kqq(pd,ma,t,it,hs,p,q,fp,yin,tt,um,xm,lm,ps);
tt=tz(1,:); fp=tz(2,:);
t=t+0.5*hs;p=1;q=1;
tz=kqq(pd,ma,t,it,hs,p,q,fp,yin,tt,um,xm,lm,ps);

tt=tz(1,:); fp=tz(2,:);
it=it+sign(hs);
y1=yin+tt/6;

function tz=kqq(pd,ma,t,it,hs,p,q,fp,yin,tt,um,xm,lm,ps)
% Runge-Kutta : get function values

z=yin+p*hs*fp;
ff=feval(pd,t,it,z,yin,hs,um,xm,lm,ps);
fp=ff;
tz=[tt+q*hs*fp;fp] ;



function xk=lqq(pd,nx,nu,nn,fil,ma,t,it,hs,um,xm,lm,ps)
%Runge-Kutta for adjoint differential equation
 (time reversed)

yin=fil(nn+1,:);xk(nn+1,:)=yin;

while it > 0 %< nn+1
[y2,it2,t2]=rqq(pd,ma,t,it,hs,yin,nx,nu,nn,um,xm,lm,ps);
xk(it,:)=y2;
it=it2; t=t2;
yin=y2;
end



function ix=iqq(vv,hh,xi)
% Linear interpolation (forwards)
nn=length(vv')-1;
nz=1/hh;
fr=nz*mod(xi,hh);
bb=floor(xi*nz)+1;

if bb <= nn, vu=vv(bb+1,:);
else vu=vv(nn);
end

ix=(1-fr)*vv(bb,:) + fr*vu;


function ix=jqq(vv,hh,xi)
% Linear interpolation (backwards)
```

APPENDIX D: SCOM Package

```
global um xm jm
nn=length(vv')-1;
nz=-1/hh;

if xi < 0, xi = 0;
end
fr=-nz*mod(xi,hh);
bb=ceil(xi*nz)+1;

if bb>1, vw=vv(bb-1,:);
else vw=vv(1);
end

ix=(1-fr)*vv(bb,:) + fr*vw;
```

Note that linear interpolation of the state is required in integrating the objective function, and in solving the adjoint differential equation. The latter is solved backwards in time t, so needs backwards interpolation.

Appendix E
Format of Problem Optimization

The control function approximates the control function $u(.)$ by a step-function, dividing $[0, 1]$ into nn equal subintervals. (Because the dynamic equation has a smoothing effect, set function controls are usually a sufficient approximation.) Function values (and often also gradients with respect to control) for the objective function are obtained by solving differential equations. They are then supplied to the optimization program constr in MATLAB's Optimization Toolbox. The computation is considerably faster if gradients are supplied, but this is not suitable for some problems, especially if the functions are not well behaved outside the feasible region of the problem. If gradients are used, then the adjoint differential equation is required. If a constraint $x(1) = k$ is required, then it is added to s a positive penalty parameter. The user must supply a calling program (defining all parameters), and user subroutines for the functions of the given control problem. This use of MATLAB minimizes the amount of programming required for a given control problem, since MATLAB handles matrix calculations and passing of parameters very effectively. (But another programming language may be preferred for a large control problem, if faster computation is needed.) The acronym SCOM stands for step-function control optimization on Macintosh, since the intention was to compute optimal control on a desktop computer, rather than on a mainframe or workstation. Note that MATLAB on a Windows computer does the same job.

Appendix F
A Sample Test Problem

The following example has two controls and one state,
and gradients are calculated.

```
subs=cell(1,9);
subs={'t3x','t3j','t3f','t3c','t3k','t3l','t3g','t3a','t3p'};
par=[1, 2, 20, 0, 1, 0.25]; % Parameters
% nx, nu, nn, npa, grad, c

nnb=par(3); nx=par(1); nu=par(2);
xinit=[0.5];
% Initial value for the state

u0=zeros(nn,nu);
% Starting values for computing the control

ul=zeros(nn,nu);
% Lower bound (vector or matrix) for the control

uu=ones(nn,nu);
% Upper bound (vector or matrix) for the control

figure


Control=constr('fqq',u0,[],ul,uu,'gqq',par,subs,xinit)
% Calls constr package

[Objective,Constraint,State,Integral,Costate,Gradient]=
cqq(Control,par,subs,xinit) % Right hand side of differential
equation

function yy=t3x(t,it,z,yin,hs,um,xm,lm,ps)
```

```
yy(1)=z(1)*um(floor(it),1)+um(floor(it),2);
% Dependent variables on the right hand side
  of a differential equation are coded as
  z(1), z(2), etc.

function ff=t3j(t,it,z,yin,hs,um,xm,lm,ps)
% Integrand of objective function

c=ps(6)
ff(1)=(um(floor(it),1)-1)*li3(xm,hs,t);
ff(1)=ff(1)+c*um(floor(it),2);

function ff=t3f(xf,um,xm,ps)
% Endpoint term

ff(1)=0; % Control constraint

function gg=t3c(ii,hs,um,xm,lm,ps)
gg=um(ii,1) + um(ii,1) - 1;

function dg=t3k(ii,hs,um,xm,lm,ps)
% Gradient of constraint

dg=[eye(nn);eye(nn)];

function yy=t3g(t,hs,um,xm,lm,nn)
% Gradient of objective

temp= 0.5*(lm(t,1)+lm(t+1,1));
t2=t/nn;
yy=[(1+temp)*li3(xm,hs,t2), 0.25+temp];

function yy=t3a(nn,xf,um,xm,ps)
%Boundary condition for adjoint equation (at t=1)

yy=0;

function yy=t3l(t,it,z,yin,hs,um,xm,lm,ps)
% Right hand side of adjoint equation

yy=-(1+z(1))*um(floor(it),1)+1;
```

References

[1] Ahmed, N.U. (1988). *Elements of Finite-Dimensional Systems and Control Theory*. Longman Scientific and Technical, Harlow, Essex, England.

[2] Blatt, John M. (1976). Optimal control with a cost of switching control. *J. Austral. Math. Soc.*, 19:316-332.

[3] Brekke, K.A. and Øksendal, B. (1991). A verification theorem for combined stochastic control and impulse control. *Stochastic Anal. Related Topics*, 6:211-220.

[4] Brigham, E. and Houston, J. (2000) *Fundamentals of Financial Management*, 9th edn. The Dryden Press, Harcourt Brace College Publishers, Orlando.

[5] Bryson, A.E.Jr. and Ho, Y.G. (1975). *Applied Optimal Control*. Halsted Press, New York.

[6] Cadenillas, A. and Zapatero, F. (2000). Classical and impulse stochastic control of the exchange rate using interest rates and reserves. *Math. Finance*, 10:141-156.

[7] Campbell, J. Y., Lo, A. W. and MacKinlay, A. C. (1997). *The Econometrics of Financial Markets*. Princeton University Press, Princeton, New Jersey.

[8] Cesari, L. (1983). *Optimization - Theory and Applications*. Springer-Verlag, New York.

[9] Chakravarty, S. (1969). *Capital and Economic Development Planning*. MIT Press, Cambridge.

[10] Chen, P. and Craven, B.D. (2002). Computing switching times in bang-bang control. Mimeo, University of Melbourne.

[11] Chen, P. and Islam, S.M.N. (2002). Optimal financing for corporations: optimal control with switching time and computational experiments using CSTVA. Paper presented at the Financial Modeling seminar, Victoria University of Technology, Melbourne, Australia.

[12] Clarke, C.W. (1976). *Mathematical Bioeconomics: The Optimal Management of Renewable Resources*. John Wiley, New York.

[13] Craven, B.D. (1978). *Mathematical Programming and Control Theory*. Chapman and Hall, London.

[14] Craven, B.D. (1995). *Control and Optimization*. Chapman and Hall Mathematics, London.

[15] Craven, B.D., HAAS, K.De. and Wettenhall, J.M. (1998). Computing optimal control. *Dynamics of Continuous, Discrete and Impulsive Systems*, pp. 601-615.

[16] Craven, B.D. (1999). Computing optimal control on MATLAB, Optimization Day. Mimeo, University of Ballarat, Melbourne, Australia.

[17] Craven, B.D. (1999). Optimal control for an obstruction problem. *Journal of Optimization Theory and Applications*, 100.

[18] Craven, B.D. and Islam, S.M.N. (2001). Computing optimal control on Matlab - The SCOM package and economic growth models. *Optimization and Related Topics*, 61-70, Kluwer Academic Publishers, Amsterdam.

[19] Cuthbertson, K. (1997). *Quantitative Financial Economics, Stocks, Bonds and Foreign Exchange*. John Wiley and Sons Ltd., West Sussex.

[20] Dadebo, S.A., McCauley, K.B. and McLellan, P.J. (1998). On the computation of optimal singular and bang-bang controls. *Optimal Control Applications and Methods*, 19:287-297.

[21] Dasgupta, S. and Titman, S. (1998). Pricing strategy and financial policy. *The Review of Financial Studies*, 11:705-737.

[22] Davis, B.E. and Elzinga, D. Jack. (1970). The solution of an optimal control problem in financial modeling. *Operations Research*, 19:1419-1433.

[23] Davis, B.E. (1970). Investment and rate of return for the regulated firm. *The Bell Journal of Economics and Management Science*, 1:245-270.

[24] Dolezal, J. (1981). On the solution of optimal control problems involving parameters and general boundary conditions, *Kybernetika*, 17:71-81.

[25] Eatwell, J., Milgate, M. and Nuemann, P. (1989). *The New Palgrave Dictionary of Finance*. MacMillan Press, London.

[26] Elton, E. and Gruber, M. (1975). *Finance as a Dynamic Process*. Prentice-Hall, Englewood Cliffs, NJ.

[27] Evans, L.C. and Friedman, A. (1979). Optimal stochastic switching and the Dirichlet problem for the bellman equation, *Trans, Amer. Math. Soc.*, 253:365-389.

[28] Fletcher, R. (1980). *Practical Methods of Optimization. Vol.1. Unconstrained Optimization*. Wiley-Interscience Publication, New York.

[29] Fleming, W.H. and Rishel, R.W. (1975). *Deterministic and Stochastic Optimal Control*. Application of Mathematics, No.1. Springer-Verlag, Berlin-New York.

[30] Fröberg, Carl-Erick. (1965). *Introduction to Numerical Analysis*. Addison-Wesley Publishing Company, Reading, Massachusetts.

[31] Garrad, W.L. and Jordan, J.M. (1977). Design of nonlinear automatic flight control system. *Automatica*, 19:497-505.

[32] Giannessi, F. (1996). Private Communication. University of Pisa, Pisa, Italy.

REFERENCES

[33] Goh, G.J. and Teo, K.L. (1987). *MISER, an Optimal Control Software.* Department of Industrial and Systems Engineering, National University of Singapore.

[34] Hasdorff.L. (1976). *Gradient Optimization and Nonlinear Control.* John Wiley and Sons, New York.

[35] Isaacs, R. (1965). *Differential Games.* John Wiley, New York.

[36] Islam, S.M.N. (2001). *Optimal Growth Economics: An Investigation of the Contemporary Issues, and Sustainability Implications,* Contributions to Economic Analysis. North Holland Publishing, Amsterdam.

[37] Islam, S. M. N. and Craven B.D. (2001). Computation of non-Linear continuous optimal growth models: Experiments with Optimal control algorithms and computer programs, *Economic Modeling.* 18:551-586.

[38] Islam, S. M. N. and Craven B.D. (2002). *Dynamic optimization models in finance: some extensions to the framework, models, and computation.* Research Monograph, CSES, Victoria University, Melbourne.

[39] Islam, S. M. N. and Oh, K.B. (2003). *Applied Financial Econometrics in E-Commerce,* Series Contributions to Economic Analysis. North Holland Publishing, Amsterdam.

[40] Jennings, L.S., Fisher, M.E., Teo, K.L. and Goh, G.J. (1991). MISER3.0: Solving optimal control problems - an update. *Advances in Engineering Software,* 13.

[41] Jennings, L.S. and Teo, K.L. (1990). A numerical algorithm for constrained optimal control problem with applications to harvesting. *Dynamics of Complex Interconnected Biological Systems,* 218-234.

[42] Jennings, L.S. and Teo, K.L. (1991). A computational algorithm for functional inequality constrained optimization problems. *Automatica,* 26:371-376.

[43] Kaya, C.Y. and Noakes, J.L. (1994). A global control law with implications in time optimal control. In Proceedings of the 33rd IEEE Conference on Decision and Control, pp. 3823-3824, Orlando, Florida.

[44] Kaya, C.Y. and Noakes, J.L. (1996). Computations and time-optimal controls. Optimal Control Applications and Methods, 17:171-185.

[45] Kaya, C.Y. and Noakes, J.L. (1997). Geodesics and an optimal control algorithm. In Proceedings of the 36th IEEE, pp. 4918-4919, San Diego, California.

[46] Kaya, C.Y. and Noakes, J.L. (1998). The leap-frog algorithm and optimal control: Background and demonstration. In Proceedings of International Conference on Optimization Techniques and Applications (ICOTA '98), pp. 835-842, Perth, Australia.

[47] Kaya, C.Y. and Noakes, J.L. (1998). The leap-frog algorithm and optimal control: theoretical aspects. In Proceedings of International Conference on Optimization Techniques and Applications (ICOTA '98), pp. 843-850, Perth, Australia.

[48] Kendrick, D.A. and Taylor, L. (1971). *Numerical Methods and Nonlinear Optimizing Models for Economic Planning.* Studies in Development Planning, Cambridge, Mass.

[49] Krouse, C.G. and Lee, W.Y. (1973). Optimal equity financing of the corporation. *Journal of Financial and Quantitative Analysis*, 8:539-563.

[50] Lee, E.B. and Markus, L. (1967). *Foundations of Optimal Control Theory*. John Wiley and Sons, New York.

[51] Lee, H.W.J., Teo, K.L., Rehbock, V. and Jennings, L.S. (1997). Control parameterization enhancing technique for time optimal control problems. *Dynamic Systems and Applications*, 6:243-262.

[52] Leonard, D. and Long, N.V. (1992). *Optimal Control Theory and Static Optimization in Economics*. Cambridge University Press, Melbourne.

[53] Li, X. and Yong, J. (1995). *Optimal Control Theory for Infinite Dimensional Systems*. Birkauser Boston, Boston.

[54] Liu, L. and Teo, K.L. (2000). Computational Method For a Class of Optimal Switching Control Problems. *Progress in Optimization*, 221-237.

[55] Matula, J. (1987). An extreme problem. *Journal of Australian Mathematical Society*, 28:376-392.

[56] Miele, J. (1975). Recent advances in gradient algorithms for optimal control problems. *Journal of Optimization Theory and Applications*, 17:361-430.

[57] Miele, A. and Wang, T. (1986). Primal-dual properties of sequential gradient-restoration algorithms for optimal control problems, Part 1, Basic Problem. *Integral Methods in Science and Engineering*, 577-607.

[58] Miele, A. and Wang, T. (1986). Primal-dual properties of sequential gradient-restoration algorithms for optimal control problems, Part 2, General Problem. *Journal of Mathematical Analysis and Applications*, 119:21-54.

[59] Modigliani, F. and Miller, M.H. (1958). The cost of capital, corporation finance, and the theory of investment. *American Economic Review*, 48:261-97.

[60] Modigliani, F. and Miller, M.H. (1963). Corporation income taxes and the cost of capital. *American Economic Review*, 53:433-43.

[61] Morellec, E. (2001). Asset liquidity, capital structure, and secured debt. *Journal of Financial Economics*, 61:173-206.

[62] Mundaca, G. and Øksendal, B. (1998). Optimal stochastic intervention control with application to the exchange rate. *J.Math. Economy*, 29:225-243.

[63] Nerlove, M. and Arrow, K.J. (1962). Optimal advertising policy under dynamic conditions. *Economica*, 39:129-142.

[64] Noakes, J. Lyle (1997). A global algorithm for geodesics. *Journal of the Australian Mathematical Society*, Series A, 37-50.

[65] Nordecai Avriel (1976). *Nonlinear Programming Analysis and Methods*. Technion-Israel Institute of Technology, Haifa, Israel.

REFERENCES

[66] Noussair, E.S. (1977). On the existence of piecewise continuous optimal controls. *Journal of Australia Mathematical Society*, Series B, 20:31-37.

[67] Oh, K.B. and Islam, S.M.N. (2001). *Empirical Finance of E-Commerce: A Quantitative Study of the Financial Issues of the Knowledge Economy*. CSES Research Monograph, Victoria University, Melbourne.

[68] Perthame, B. (1984). Continuous and impulse control of diffusion process in R^N. *Nonlinear Anal*, 8:1227-1239.

[69] Pontryagin, L.S., Boltyanskii, V.G., Gamkrelidze, R.V. and Mishchenko, E.F. (1962). *The Mathematical Theory of Optimal Processes*. John Wiley, New York.

[70] Powell, M.J.D. Private communication.

[71] Redbock, V., Teo, K.L. and Jennings, L.S. (1994). Suboptimal feedback control for a class of nonlinear systems using spline interpolation. *Discrete and Continuous Dynamical Systems*, 1:223-236.

[72] Redbock V., Teo, K.L., Jennings, L.S. and IEE, H.W.J. (1999). A survey of the control parameterization and control parameterization enhancing methods for constrained optimal control problems. *Progress in Optimization*, 247-275.

[73] Richard, R. Lumley and Mihall Zervos (2001). A model for investments in the material resource industry with switching costs. *Mathematics of Operations Research*, 26:637-653.

[74] Rockafellar, R.T. (1974). Lagrange multiplier functions and duality in non-convex programming. *SIAM J. Control*, 12:268-287.

[75] Sakama, Y. and Shindo, Y. (1980). On global convergence of an algorithm for optimal control. *IEEE Transaction Automatic Control*, AC-25:1149-1153.

[76] Sakama, Y. (1981). On local convergence of an algorithm for optimal control. *Numerical Functional Analysis and Optimization*, 3:301-319.

[77] Schwartz, A., Polak, E. and Chen Y. (1997). *Recursive Integration Optimal Trajectory Solver 95. A MATLAB TOOLBOX for solving Optimal control problems*, Version 1.0, Stanford University, California.

[78] Sethi, S.P. (1978). Optimal equity financing model of Krouse and Lee: corrections and extensions. *Journal of Financial and Quantitative Analysis*, 13:487-505.

[79] Sethi, S.P. and Thompson, G.L. (2000). *Optimal Control Theory*. Kluwer Academic Publishers, Amsterdam.

[80] Sengupta, J.K. and Fanchon, P. (1997). *Control Theory Methods in Economics*. Kluwer Academic Publishers, Boston.

[81] Tapiero, C.S. (1998). *Applied Stochastic Models and Control for Insurance and Finance*. Kluwer Academic Publishers, London.

[82] Teo, K.L. (1992). A computational approach to an optimal control problem with a cost on changing control. *Optimization*, 1:397-413.

[83] Teo, K.L. and Goh, C.J. (1988). On constrained optimization problems with nonsmooth cost functionals. *Applied Mathematics and Optimization*, 18:181-190.

[84] Teo, K.L., Goh, G.J. and Wong, K.H. (1991). *A Unified Computational Approach to Optimal Control Problems*. Longman Scientific and Technical, London.

[85] Teo, K.L. and Jennings, G.J. (1991). Optimal control with a cost on changing control. *Journal of Optimization Theory and Applications*, 68:335-357.

[86] Vidale, M.L. and Wolfe, H.B. (1957). An operations research study of sales response to advertising. *Operations Research*, 5:370-81.

[87] Yong, J. (1989). Systems governed by ordinary differential equations with continuous, switching and impulse controls. *Appl. Math.Optim.*, 20:223-235.

[88] Yong, J. (1991). Existence of the value for a differential game with switching strategies in a Banach space. *System Sci. Math. Sci.*, 4:321-340.

[89] Ziemba, W.T. and Vickson, R.G. (1975). *Stochastic Optimization Models in Finance*. Academic Press, New York.

Index

admissible controllers, 95
aggregate dynamic financial system, 40
analytical solution, xv, 1, 91, 104–108, 143
approximate solution methods, 10
approximation methods, 11
approximation problem, 12
associated problems, 6, 7
augmented Lagrangian algorithm, 11, 15, 98

bang-bang control, 1, 4, 7, 16, 17, 19, 26, 41, 47, 97, 101
bang-bang optimal control problems, 26, 27, 143
bang-bang optimal control solution, 40
big subintervals, 42–44, 101
boundary conditions, 6, 97
business cycle instabilities, 39

candidate selection systems, 13
cash management, 4
co-efficient, 7, 13
co-state function, 6, 9, 15
computational algorithms, 1, 10, 14, 20, 21, 101
computational experiments, xvii, 36, 89
computational methods, 1, 11, 19, 24, 27, 40, 97, 100, 101, 141
CONMIN, 11
constant-input arcs, 16
constr, 15, 20, 25–28, 44, 45, 102, 141, 189
continuous optimal control models, 10
control, 2, 12–14, 20, 22, 26, 27, 30, 31, 33, 36, 37, 39–47, 54, 61, 67, 86, 88, 103–105, 111–113, 115–122, 142
control discretization approach, 10
control function, xvi, 9, 14, 18, 19, 21, 23, 39, 142, 189
control mechanism, 89
control parameterization knot points, 16
control parameterization technique, 11, 12

control policy, 10, 29, 46, 54, 61, 67, 88, 101, 103, 109
control restraint, 95
control strategy, 16
control system, 11, 17
control variables, 93, 139
convergence properties, 12
convex function, 96
cost analysis, 10
cost function, 2, 11–13, 19, 27, 28, 30, 35, 36, 41, 45, 91, 142
cost of changing control, 8, 19, 20, 27, 30, 36, 41, 86, 89, 142
CPET, 16, 19
CSTVA, 19, 139, 141

damped oscillator, 12, 39, 40, 54, 143
damping function, 89
DE solver, 15, 21, 22, 26
decision variables, 13
deterministic, 1, 4, 18
deterministic optimization, 2
difference of optimal control policy sequences, 10
discount factor, 24
discrete optimal control models, 10
discretization augment, 18
division of the time intervals, 10
Dodgem problem, 12
dynamic behavior, 40
dynamic equation, 7, 18, 23, 28, 29, 102, 189
dynamic financial system, 1, 39, 40, 142
dynamic optimization models, 1, 3
dynamic programming, 10, 11

economic models, 14, 15
end-point constraint, 98
end-points, 10, 42, 43

financial decision making, 3, 30, 39, 41

financial engineering, 4
financial market, 89, 92
financial optimal control models, 2, 8, 18, 22, 139, 141–143
financial optimization models, 2, 4
financial planning, 4, 142
financial sector, 39, 89
financial system, 1, 39–41, 43, 54, 88, 102, 111, 115, 139, 142
finite difference approximation, 15
finite difference method, 141
finite horizon models, 4
Fish problem, 12
fishery harvesting model, 12
fitting function, 21, 30–33, 35, 48, 54, 109, 110, 113–115, 118–122, 139
fixed-time control, 3
fixed-time optimal control problem, 11, 23, 25
fixed-time period, 19
forcing function, 41, 114

global minimum, 5, 10, 18, 54, 116, 122
gradient search methods, 10, 11
grid-points, 11, 14, 19, 26, 29, 31, 45, 102

Hamiltonian, 4, 6, 9, 15, 96

implicit constraints, 14, 15
indifference principle, 8–10
infinite horizon models, 4
initial and terminal conditions, 4
inter-temporal time preference, 3
interpolation, 11
investment, 4, 30, 36, 41, 89, 91–93, 102, 110, 113, 141–143
investment allocation, 91, 143
investment model, 16

Karush-Kuhn-Tucker, 5

Lagrange multipliers, 5, 98, 99
Leap-Frog Algorithm, 17
linear interpolation, 11, 15, 29, 46, 103, 187
local minimum, xvii, 4, 10, 116, 118, 122
low-pass filter, 11

mathematical programming, 4, 94
mathematical structure, 1, 4
MATLAB, xvi, xvii, 11, 12, 14, 15, 18, 20, 25–28, 44, 45, 101, 102, 141, 189
Maximum Principle, 2, 4, 94, 96
maximum principle, 8–10
minimization problem, 2, 4–6, 18, 27, 30, 44, 98, 102
MISER, 11, 12, 14, 17–19
mixed discrete and continuous optimal control problem, 13

multi-state functions, 40

necessary conditions, 6, 17, 96
non-linear complex dynamic behavior, 89
non-linear control system, 17
non-linear transformation, 20, 21
non-smooth functions, 12
nqq computer package, 41
nqq function, 20

obstacle, 18
OCIM, 11, 12, 14, 18
ODE, 15, 19
operations research, 2
optimal consumption, 4
optimal control modeling, 1, 142
optimal control problems, 1, 2, 4, 7, 8, 10–15, 17, 20, 23, 25
optimal control theory, 1
optimal controller, 95
optimal corporate finance, 4, 92
optimal investment planning, 4, 20, 36, 89, 141
optimal investment strategy, 89
optimal path, 18
optimal portfolio choice, 4
optimization criteria, 3
oscillator problem, 12, 36, 37, 40, 44–46, 120, 124, 142
oscillatory behavior, 39
oscillatory dynamics, 39, 89
oscillatory financial model, 39

penalty constant, 13
penalty term, 14, 91, 98, 99, 142
piece-wise continuous function, 2
piecewise-linear transformation, 20, 21, 25, 29, 141
planning horizon, 2, 6, 31, 94, 98, 101
Pontryagin theorem, 2, 6, 8, 9, 18
Pontryagin's Maximum Principle, 17

Quadratic Programming sub-problem, 25
quasi-Newton updating method, 11, 25

real-world problems, 1, 2
restricted domain, 18
reverse-time construction technique, 94
RIOTS, 11, 12, 18, 19
risk management, 4

scaled time, 22, 28–30, 42, 43, 45–47, 100, 102, 103
SCOM, 12, 21, 25
second order differential equation, 20, 39, 40, 45, 46, 141
sequential quadratic programming method, SQP, 20, 25, 141
single-input non-linear system, 16

INDEX

singular arc, 7, 16, 26, 41, 97
stabilization mechanism, 39
stable optimal investment planning, 89
state variables, 93, 97, 111
STC, 16, 17, 19
step function, 3, 10, 11, 14, 15, 18–21, 25, 26, 40, 120, 141, 142, 189
stochastic, 1, 4, 16, 18
stochastic finance, 4
stochastic financial optimal control model, 18
STV algorithm, 19, 20, 141, 143
switching control, 3, 27, 41, 86, 89
switching time computation method, 16
switching time variable method, 141
switching times, 13, 17, 19, 26, 27, 29, 31, 32, 35, 36, 39, 43, 46, 47, 67, 89

system parameters, 13, 44

terminal constraint transformation, 98
terminal manifold, 94
time optimal control problems, 16
time scale, 23, 39, 41, 42, 99
time subdivision, 40, 108
time transformation, 21, 101
time-optimal control, 3
time-optimal control problems, 23
TOBC, 17
TPBVP, 17
two-point boundary-value problem, 17

valuation, 4

DISCARDED
CONCORDIA UNIV. LIBRARY
CONCORDIA UNIVERSITY LIBRARIES
SIR GEORGE WILLIAMS CAMPUS
WEBSTER LIBRARY